1971. 1st Edition. Nice Copy £10-00

CAB 127

MESSERSCHMITT

MESSERSCHMITT

an aircraft album

J. RICHARD SMITH

LONDON IAN ALLAN

First published 1971

SBN 7110 0224 X

Published by Ian Allan Ltd, Shepperton, Surrey and printed in the United Kingdom by Morrison and Gibb Ltd, London and Edinburgh

PHOTOGRAPH CREDITS

We are grateful to the following for providing the photographs on these pages:
Eddie Creek 74, 88, 99, 109; Imperial War Museum 9, 58; Zezvonil Jeroslav via Eddie Creek 134; Messerschmitt 11, 105; Messerschmitt via Eddie Creek 83, 103, 107; RAF 12, Harold Thiele via Eddie Creek 69 (bottom).
All of the other photographs except those on pages 6 and 108 were supplied by Heinz J. Nowarra.

Contents

Introduction

WILLY MESSERSCHMITT was born on 26 June 1898, the son of a Frankfurt wine merchant. He first became interested in aviation at the age of eleven after reading about the crash of a Zeppelin airship at Echterdingen in the newspapers. During the summer of 1909 he was occasionally lucky enough to see a Zeppelin at Freidrichshafen and also managed a visit to the Internationale Luftfahrt-Ausstellung (International Aviation Exhibition) in Frankfurt. He was so excited by these events that he began to build his own model aircraft.

In 1910, Messerschmitt's parents moved to Bamberg, a beautiful old town in Franken, some 30 miles north of Nuremberg. It was here that Messerschmitt made the acquaintance of Freidrich Harth, an architect who had pioneered gliding flight. He was much impressed by Harth's work, and soon afterwards made his first glider flight from the Ludwager Kulm, a hill near Bamberg.

When Harth tested his fourth glider, Messerschmitt, who was then 15 years old, helped him but they found the area around Bamberg was not particularly suitable for such flying. In the summer of 1914, the friends began a move to the Heidelstein in the Rhoen, but the declaration of the First World War led to Harth, who was then 34, joining the German army. During his absence, Messerschmitt completed the construction of a glider which Harth had designed, and they successfully flew it in 1915 when Harth returned from the front.

In 1916 the first jointly designed Harth-Messerschmitt glider was built. It was a high wing monoplane with an uncovered fuselage built of steel tube, the pilot sitting under the wing leading edge. The aircraft had a wing span of 39 ft 4½ ins and utilised a wing-warping system of control. During a test flight made in a wind speed of 40 ft/sec, the aircraft attained an altitude of about 30 feet and stayed in the air for 3½ minutes.

Because of his experience in aviation, Harth was transferred, in 1917, to the Flieger-Ersatz-Abteilung (Pilot Replacement Section) at Schleissheim. Here he served as an instructor on aircraft bracing techniques. In the same year, Messerschmitt joined the army, serving for a short time with a minewerfer unit. He was then transferred to the Schleissheim school which has often been described as the 'cradle of Bavarian aviation'.

After the armistice in November 1918, Harth joined the Bayerische Flugzeug-Werke (Bavarian Aircraft Company) a subsidiary of the famous Albatros works, and Messerschmitt was enrolled as a student at the Technical High School in Munich. Although aviation development was severely restricted in Germany, glider flying continued unchecked. Oskar Ursinus, who had designed a remarkable fighter seaplane with retractable floats during the

Facing page: *The most famous Messerschmitt design, here represented by a Bf 109E-4 of 2./JG 27 in the Middle East in 1941.*

One of the first successful designs produced by the Harth-Messerschmitt partnership was the S 9 glider of 1920.

First World War, called for all young airmen to make the Rhoen area the centre for all future glider building and flying. Harth and Messerschmitt were among the first to answer his call.

Two years after the end of the 1914–18 war the first annual glider competition took place in the Rhoen at the Wasserkuppe, but it was not until two years later that Harth and Messerschmitt moved to the area, preferring until then to stay at the Heidelstein, some five miles to the south-east. During a test flight in the Pilotus glider early in 1921, Harth crashed, receiving severe concussion and a fractured pelvis. He never completely recovered from these injuries. Undismayed, Messerschmitt continued development of the glider under the designation S 10. This machine was successfully flown by Wolf Hirth, who later became the most famous glider pilot in pre-Nazi Germany.

During the early 1920s, Harth-Messerschmitt designs gained a growing reputation, although some doubts were cast on the structural strength of the tail units of their aircraft. After breaking with Harth in 1923, Messerschmitt founded his own group at Bamberg known as the Messerschmitt-Flugzeugbau. Despite the founding of the new group, Messerschmitt remained

Top: *The last aircraft to be designed and built by the Bayerische Flugzeug-werke was the BFW 3—a development of the Udet U 12.*

Left: *The Messerschmitt M 23b lightplane was one of the earliest of a series of low-wing monoplanes which was eventually to culminate in the Bf 109.*

Facing page: *A Messerschmitt of Bf 108B Taifun of the* Luftwaffe. *An excellent machine in its own right, the Bf 108 was the immediate progenitor of the Bf 109.*

at the Technical High School, being the first student to be allowed to build a glider as the major part of his thesis, He was so successful in this that he passed his examinations with flying colours and became a fully-qualified engineer.

Perhaps the most important influence on Messerschmitt at this time was provided by Theo Cronweiss, director of Sportflug GmbH. It was Cronweiss who encouraged the design of both the M 17 lightplane and the M 18 airliner. This excellent and very economical machine was designed to be built at a price not exceeding 25,000 RM (then about £1,250 Sterling). Its success led to the expansion of the group to become the Messerschmitt-Flugzeugbau GmbH and the formation of the Nordbayerische Verkehrsflug airline from 50,000 RM capital provided by it and Sportflug GmbH.

The success of the Messerschmitt M 18 led its designer to request a subsidy from the Bavarian government, but as that body was already heavily committed to financing the BFW company it was unable to help. However a suggestion was made that the two companies should merge, BFW undertaking to produce only Messerschmitt aircraft whilst Messerschmitt gave first priority to BFW for the production of his aircraft. Eventually an agreement was signed on 8 September 1927 and Messerschmitt became chief designer of the united companies with his headquarters in Augsburg.

Following the failure of the first M 20 airliner, Messerschmitt persuaded the Strohmeyer-Raulino family to buy, in July 1928, the shares in BFW owned by the Bavarian government and the Reichs-Verkehrs-Ministerium. It should be mentioned here that Frau Messerschmitt's

maiden name was Fraulein Raulino, his wife playing no small part in the early development of the company. Following the take over of BFW, Messerschmitt became joint managing director with Fritz Hille.

Despite the success of the M 18 and an order for twelve M 20s, the constant design and development of new types did little to consolidate the financial position of the Bayerische Flugzeugwerke AG. In June 1931 the company finally went bankrupt. Messerschmitt managed, by dint of selling his car and negotiating with a Rumanian company for the licence production of the M 23b, to retain a small group of men under the name of Messerschmitt-Flugzeugbau GmbH. During the year that followed, an attempt was made to reimburse all creditors with some success. The only real commercial success at this time was the M 23 lightplane, one of which was bought by Rudolf Hess.

Thus began a long-standing friendship between the deputy leader of the Nazi party and Messerschmitt, a friendship which was to save his Augsburg factory from being taken over in 1932 for use as a municipal tram depot. When the Nazis swept to power in 1933, Messerschmitt was one of the first to be honoured, the remnants of his company forming the basis of a newly constituted Bayerische Flugzeugwerke AG with 82 employees. The company also took over a newly-completed factory at Regensburg.

In one way the establishment of the Nazi Party was a mixed blessing to Messerschmitt. His long-standing enemy, Erhard Milch was appointed as Secretary of State for Air, and he decided that the new BFW company should be limited solely to the licence production of other manufacturers' aircraft. This the management were forced to accept, so in 1933, BFW were given an order for ten Heinkel He 45c reconnaissance aircraft. Ernst Heinkel had previously tried to buy BFW's Augsburg factory, and was not well disposed towards Messerschmitt, so the situation developed where he refused to let BFW's technicians inspect the aircraft they were supposed to be producing!

In desperation, Messerschmitt decided to seek orders from abroad, and it was this that eventually resulted in an order from the RLM for a single-seat fighter, the Bf 109. Meanwhile the company produced its first all-metal aircraft, the Bf 108 Taifun which gained many international records. But it was the Bf 109 that really made Messerschmitt's name. First flown in September 1935, the Bf 109 was under constant development during the war in Germany and after it in Czechoslovakia and Spain. Even today if the name of Messerschmitt is mentioned it brings forth a vision of the small, somewhat angular single-seat fighter.

Eight months after the Bf 109 took to the air, the Bf 110 twin-engined destroyer made its first flight. Designed to a very interesting concept, the Bf 110 proved a dismal failure during the Battle of Britain, but nevertheless provided the *Luftwaffe* with its most important night fighter for the greater part of the war. The last three aircraft designed for the Bayerische Flugzeugwerke were the Bf 161 and Bf 162 Jaguar, both based on Bf 110 and the Bf 163 short take-off reconnaissance project.

On 11 July 1938, the Bayerische Flugzeugwerke AG was renamed as the Messerschmitt AG, the company retaining this title until the end of the war. The factories were now building three aircraft, the Bf 108, Bf 109 and Bf 110, and despite the expansion of the Augsburg and Regensburg plants, it was realised that these would never be big enough to cope with the huge quantities ordered by the RLM or German Air Ministry. It was therefore decided that other companies should undertake licence production of the Bf 109. These included Ago at Oschersleben, Erla at Leipzig (who eventually produced more Bf 109s than any other company), Fieseler at Kassel, Arado at Brandenburg and WNF at Wiener-Neustadt. It is perhaps worth mentioning here that Messerschmitt became the largest German aircraft manufacturer during the Second World War.

The first aircraft to receive an 'Me' designation was the Messerschmitt Me 209 V1 which held the world speed record for piston-engined aircraft for over 30 years. This was followed by the Me 210 which has been described as one of the most dismal failures produced by the German aircraft industry. This was eventually developed into the greatly-improved Me 410 Hornisse. The remarkable Me 261 Adolphine was quite successful and the Me 264 was one of the first attempts to produce an inter-continental bomber.

Of Messerschmitt's later designs, undoubtedly the most famous was the Me 262 —the world's first operational jet-propelled fighter. This aircraft more than any other could have changed the course of the war in Germany's favour, but Hitler's insistence

Although the Messerschmitt Bf 110C–4 proved a dismal failure during the Battle of Britain, it was one of the Luftwaffe*'s most successful night fighters.*

that it be used as a bomber delayed its introduction into operational service by almost a year.

Perhaps the most spectacular design to emerge from the Messerschmitt stable was the Me 163 Komet rocket-propelled fighter. The design of this aircraft owed much to the work of Dr Alexander Lippisch who joined Messerschmitt in January 1939. A brilliant innovator, it was Lippisch who conceived the concept of the delta wing, a planform now used on many of the world's most advanced aircraft.

Messerschmitt also made several attempts to develop a successor to the Bf 109 with the Me 209 V, the Me 309 and the Me 209A, but none of these aircraft were produced in any numbers. Rather untypical of the company was the huge and ugly Me 321 glider and Me 323 transport which provided the *Luftwaffe* with greatly increased cargo carrying capacity. Another rather spectacular and yet extremely crude aircraft was the Messerschmitt Me 328 pulsejet powered light bomber. It was perhaps fortunate for the *Luftwaffe* that this aircraft never saw operational service.

The ex 2.(F)/122 Me 410A-1/U1 (probably w/nr 420430) which was captured in the Middle East and allocated the Air Ministry number 72. The machine is at present at RAF Cosford.

After the war, Messerschmitt went to Spain, developing the Messerschmitt Me 100 and Me 200 Pfeil. The former, which was built as the Hispano HA-100-EI was a tandem two-seat trainer and the latter, which was also known as the Hispano HA-200-RI Saeta was also a trainer, but achieved fame for being the first jet-propelled aircraft built in Spain.

During the 1960s, the reconstituted Messerschmitt AG designed several jet transport projects, perhaps the most important being the Me 308 Jet Taifun. The company later amalgamated with Bolkow to become the Messerschmitt-Bolkow GmbH, and in 1969 merged with the Hamburger Flugzeugbau GmbH to become Messerschmitt-Bolkow-Blohm GmbH.

Currently (1970) the largest German aviation company with 19,400 employees, MBB has its headquarters in Munich. In addition to producing the F-104G Starfighter under licence, the company took over development of the HFB 320 Hansa Jet which, designed by Dipl Ing Wocke, had made its first flight on 21 April 1964. The Hansa Jet is a light airliner powered by two 2,950 lbs thrust General Electric CJ610-5 turbojets, capable of carrying up to twelve passengers. An unusual feature of the design is the *forward* swept wing, a characteristic inherited from the Ju 287 bomber of 1945. Apart from producing the Hansa Jet, MBB is also sharing in the production of the Fokker F.28 Fellowship 65-seat jet airliner of which 30 have been ordered.

Aeroplane Types

S 13 and S 14

THE SUCCESS of the S 10 glider led Harth and Messerschmitt to develop an improved version, the S 11. The main difference between the two aircraft was that the latter had all its control cables buried in the wing. For the third Rhoen competition, held in 1932, the Harth-Messerschmitt team developed the S 13 and S 14 gliders. On 13 September 1921, the S 13 with Harth at the controls, gained an endurance record by staying in the air for 21 minutes.

The Harth-Messerschmitt S 14 was a development of the S 13 which introduced a covered fuselage for the first time. Shortly after its first take-off, the S 14 crashed, severely damaging the pilot Hirth's larynx. A second machine also crashed but Messerschmitt seemed little concerned for the fate of the pilots, concentrating on discovering the cause of the crashes. Eventually, the S 14 was rebuilt and established a new altitude record with Hackmack at the controls.

Specification (S 14)

SPAN	14·00 m	(45 ft 11¼ ins)
LENGTH	5·40 m	(17 ft 8½ ins)
EMPTY WEIGHT	105 kg	(232 lbs)
LOADED WEIGHT	175 kg	(386 lbs)

Development of the Harth-Messerschmitt Pilotus, *seen here during an early flight, eventually led to the S 13 and S 14 gliders.*

S 15

IN 1923, Messerschmitt decided at last to turn to powered aircraft as a solution to the problems of sustained flight. This created a difference of opinion between him and Harth, and eventually they split up, Harth eventually drifting away from aviation to other interests.

Although Messerschmitt decided to switch to powered aircraft design, it should be remembered that he did a tremendous amount for the development of glider flight, He had developed one of the most successful aerofoil sections for gliders, the Göttingen 535 which was used on such famous designs as the Consul of 1923, the Fafnir of 1930, the Windspeil, Rhoenbussard and Rhönsperber of 1924–33, the Moazagotl of 1933 and the DFS Kranich of 1935. This section had been developed from the Göttingen 534 which had been designed by Harth.

At the fifth Rhoen Competition in 1924 Messerschmitt introduced the S 15 which was essentially a powered version of the S 14 glider. The aircraft was powered initially by a 10 hp Victoria engine, although this was soon replaced by a 14 hp Douglas unit. Inexperience with aero engines led to the S 15 losing its propeller and crashing.

Specification

SPAN	14·60 m	(47 ft 10½ ins)
LENGTH	5·00 m	(16 ft 5 ins)
EMPTY WEIGHT	180 kg	(397 lbs)
LOADED WEIGHT	300 kg	(662 lbs)
MAXIMUM SPEED	120 km/h	(75 mph)

S 16 Bubi

THE MESSERSCHMITT S 16 (the 'S' being an abbreviation of *Segelflugzaug* or sailplane) was the first aircraft designed solely by Messerschmitt. It was a two-seat light sporting aircraft similar in appearance to the S 15, but with a much narrower chord fin and rudder. It was powered by a 21 hp Douglas Sprite engine driving a tractor airscrew. The aircraft performed well during the 1924 Rhoen competition, making some remarkable flights with *Lt* Heinrich Seywald, a First World War ace with six victories to his credit, at the controls.

Although able to reach a maximum speed of 71 mph, the S 16 possessed the remarkably low fuel consumption of nine miles to the gallon. A modified version was produced under the designation S 16a *Betty*. Following the disappearance of Harth from the design partnership, all new Messerschmitt aircraft received the prefix 'M' to their designations. This method was retained until 1933 when the new RLM system was adopted.

Specification (S 16)

SPAN	14·40 m	(47 ft 3 ins)
EMPTY WEIGHT	218 kg	(481 lbs)
MAXIMUM SPEED	115 km/h	(71 mph)

Top: *The Harth-Messerschmitt S 15 was the first powered aircraft designed by the partnership. It competed during the fifth Rhoen Competition of 1924.*

Left: *Messerschmitt (left) posing with an unknown pilot (possibly Seywald) by the S 16. The machine also competed in the Rhoen Competition.*

M 17 Ello

EARLY IN 1924, Messerschmitt made the acquaintance of Theo Cronweiss, a former Oberleutnant with the German Air Force who had gained five victories on the Mesopotanian Front. Cronweiss, who was then a director of Sportflug GmbH, a private flying school, encouraged Messerschmitt to develop a two-seat lightplane under the designation M 17.

The aircraft was completed in 1925 and became the first really successful design to be built by the Messerschmitt company. It was a tiny high-wing monoplane, owing something to the previous S 16, but possessing, for its time, a remarkable beauty of line. The M 17 was powered by a 32 hp Bristol Cherub III engine (although it was also flown with the 24 hp ABC Scorpion). The most remarkable flight by the M 17 was when it flew over the Alps from Bamberg to Rome. The whole machine, although weighing only about 400 lbs empty could carry a considerable payload. Perhaps the best tribute accorded to the M 17 was by the *Deutsche Versuchsanstalt für Luftfahrt* (German Aviation Test Centre) who described it as 'the best airframe since 1917'.

Specification

SPAN	11·60 m	(38 ft 0¾ ins)
LENGTH	5·85 m	(19 ft 2¼ ins)
HEIGHT	1·80 m	(5 ft 11 ins)
EMPTY WEIGHT	180 kg	(397 lbs)
LOADED WEIGHT	370 kg	(816 lbs)
MAXIMUM SPEED	132 km/h	(83 mph)
SERVICE CEILING	4,000 m	(13,123 ft)

The first really successful Messerschmitt design was the tiny M 17 seen here with a Bristol Cherub engine.

M 18

The aircraft that first introduced the name of Messerschmitt to the German public was the M 18 airliner produced in 1925. The machine, which was similar in appearance to the M 17, was also built to the order of Theo Cronweiss, who as director of *Sportflug GmbH* had formed the *Nordbayerische Verkehrsflug* airline. The prototype M 18 was unregistered and featured a cabin for three passengers situated behind the open pilot's cockpit. The fuselage and wings were built of wood and fabric and the machine was powered by an 80 hp Siemens Sh 11 seven cylinder engine.

Trials were so successful that Messerschmitt requested the Bavarian government to subsidise the large-scale construction of the machine. The first production model was designated Messerschmitt M 18a and differed in being built of metal with mixed metal and fabric covering to the wing. Two aircraft were completed, D–947 (w/nr 27) *Fürth* and D–1118 (w/nr 28), both powered by Siemens Sh 11 radials. They were delivered to *Nordbayerische Verkehrsflug* in 1926, which, based at Fürth near Nuremberg, eventually operated 10 air routes to 16 different towns. These included Augsburg, Bayreuth, Chemnitz, Cottbus, Dresden, Erfurt, Gera, Halle, Hof, Heiningen, Leipzig, Plauen, Rudolstadt, Schweinefurt, Weimar and Zwickau. The aircraft were in daily use apart from Sunday. The M 18 was such an economic design that it was possible to operate at an average cost of 0·60 RM per kilometre.

The two previously mentioned aircraft were followed by a production batch of 12 Messerschmitt M 18bs which differed in being powered by 110 hp Siemens Sh 12 radials. Later aircraft were capable of carrying four passengers, and all but one were delivered to *Nordbayerische Verkehrsflug* and its successor, *Deutsche Verkehrsflug*. These included D–1133, D–1266, D–1326, D–1333, D–1347, D–1365, D–1405, D–1496, D–1567, D–1581 and D–1643. The other aircraft, D–1577 (w/nr 419) was delivered to the *Deutsches Versuchsanstalt für Luftfahrt*.

Only three Messerschmitt M 18c photo-

graphic aircraft appear to have been constructed, these being powered by 220 hp Armstrong Siddeley Lynx engines. The variant also differed from the earlier aircraft in having a lengthened undercarriage with the shock absorbing struts attached to the underside of the wings. The first aircraft, D–1860 (w/nr 448) was probably delivered to *Hansa Luftbild*—the main company in Germany to produce aerial photographs. The other two aircraft, registered CH–144 and CS–AAH were delivered to Switzerland and Portugal respectively.

The final production variant was the Messerschmitt M 18d of which eight aircraft were constructed. The type was similar in appearance to the M 18c, featuring an enclosed cockpit and provision for eight passengers. Various engines were employed including the 225 hp Armstrong Siddeley Lynx, the 150 hp Walter Mars and the 325 hp Wright Whirlwind.

Five Messerschmitt M 18ds, D–1812, D–1847, D–1848, D–1861, and D–1872 (respectively w/nrs 469, 477, 478, 479 and 480) were delivered to *Nordbayersiche Verkehrsflug AG* (later *Deutsche Verkehrsflug*. One machine registered CH–191 (w/nr 476) was operated by the Swiss Ad Astra Aero Company, later being trans-

Top: *This M 18d was probably the aircraft delivered to Hansa Luftbild and later registered D–1860.*

Right: *The Walter Mars powered Messerschmitt M 18d outside Nordbayerische Verkehrsflug's offices. This aircraft was named* Regensburg.

Facing page: *A Messerschmitt M 18b owned by Nordbayerische Verkehrsflug, registered D–1266* Schweinfurt. *This machine had a Siemens Sh 12 engine.*

ferred to Swissair as HB–IME. A seventh aircraft was fitted with floats and featured considerably increased fin area. The last machine may possibly have been delivered to the American Eastern Aircraft Company who planned to build the machine under licence. In the event these plans came to nothing following the Wall Street crash of 1929.

Specification (M 18b)

SPAN	15·60 m	(51 ft 2 ins)
LENGTH	8·05 m	(26 ft 5 ins)
HEIGHT	2·25 m	(7 ft $4\frac{3}{4}$ ins)
EMPTY WEIGHT	650 kg	(1,433 lbs)
LOADED WEIGHT	1,200 kg	(2,645 lbs)
CRUISING SPEED	140 km/h	(87 mph)
NORMAL RANGE	700 km	(435 miles)

M 19

THE FIRST low-wing monoplane designed by Messerschmitt was the M 19 single-seat lightplane. It was developed to compete in the Sachsenflug competition sponsored by the *Deutsche Versachsanstalt für Luftfahrt*. The DVL had issued stringent requirements for this competition especially as regards difference between empty weight and useful load and fuel consumption. However, when it was pointed out that the Messerschmitt M 17 already met these requirements, the DVL devised yet more difficult conditions.

Two M 19s were built by Messerschmitt, the design owing much to the M 17, also being powered by the 24 hp ABC Scorpion engine. During the technical tests the M 19 took first prize, being rated as 'infinitely good'. With Theo Cronweiss at the controls it became the overall winner of the competition, and gained for Messerschmitt a prize of 60,000 RM which he invested in the Bayerische Flugzeugwerke.

Specification

SPAN	9·60 m	(31 ft 6 ins)
LENGTH	5·40 m	(17 ft $8\frac{3}{4}$ ins)
HEIGHT	1·50 m	(4 ft $11\frac{1}{4}$ ins)
EMPTY WEIGHT	138 kg	(304 lbs)
LOADED WEIGHT	336 kg	(741 lbs)
MAXIMUM SPEED	114 km/h	(71 mph)

Top right: *The first of a long line of low-wing sporting monoplanes, the Messerschmitt M 19 won for its designer a prize of 60,000 RM.*

Right: *A redesign of the Udet U 12 biplane trainerbu ilt by the Bayerische Flugzeugwerke AG, the BFW U 12b was a most elegant machine.*

Left: *The seventh production M 18d which was fitted with floats and a Wright Whirlwind engine.*

BFW U 12b

FOLLOWING the resignation of Ernst Udet and chief-designer Scheuermann from the Munich-Ramersdorf based Udet Flugzeugbau GmbH in 1926, the company went into liquidation. In order to save it, the Bavarian government and some private sources provided enough capital to re-establish the company as the Bayerische Flugzeugwerke AG. BFW took over the hangars of the then defunct Rumpler Flugzeugwerke in Augsburg, and continued the construction of the Udet U 12 Flamingo.

The U 12a, which was an attractive biplane powered by a 80 hp Siemens Sh 11 radial, was the favourite mount of Ernst Udet during his famous aerobatic displays. Many aircraft of the type were subsequently delivered to the *Deutsche Verkehrsflieger Schule* at Schleissheim, and BFW produced an improved version known as the U 12b powered by a 108 hp Siemens Sh 12 engine. The company also developed two further modifications of the basic design, the heavily strutted BFW 1 and the BFW 3.

Specification (U 12b)

SPAN	10·00 m	(32 ft 9¾ ins)
LENGTH	7·50 m	(24 ft 7¼ ins)
HEIGHT	2·80 m	(9 ft 2¼ ins)
EMPTY WEIGHT	550 kg	(1,213 lbs)
LOADED WEIGHT	800 kg	(1,764 lbs)
MAXIMUM SPEED	140 km/h	(87 mph)

M 20

WORK BEGAN on the Messerschmitt M 20 during the autumn of 1927, the aircraft being a development of the M 18d powered by a 500 hp BMW VIa engine. The maiden flight of the prototype on 26 February 1928 was followed by disaster when fabric stripped from the trailing edge of the wing and the pilot, misjudging the seriousness of the situation baled out at 250 feet. The resulting crash led to the cancellation of the two aircraft previously ordered by *Lufthansa*.

Despite this setback, the *Bayerische Flugzeugwerke* completed a second aircraft, and this was flown quite successfully on 3 August 1928 by Theo Cronweiss. *Lufthansa* immediately reinstated their original order for the machine, the two aircraft being delivered to the airline during 1929. They were designated Messerschmitt M 20as and were registered D–1480 (w/nr 392) *Franken* and D–1676 (w/nr 421) *Schwaben*.

The success of these two aircraft led to the construction of 12 Messerschmitt M 20bs for *Lufthansa*. The new aircraft was larger, with accommodation for 10 passengers. It also introduced a modified cockpit, wing dihedral and a rounded fin and rudder assembly. The first two M 20bs, D–1928 (w/nr 442) *Rheinpfalz* and D–1930 (w/nr 443) were followed by a batch of ten aircraft, respectively allocated the w/nrs 540 to 549. These were D–2005 *Odenwald*, D–2006 *Spessart*, D–2025 *Westerwald*, D–2026 *Schwarzwald*, D–2285 *Sauerland*, D–2290 *Fläming*, D–2341 *Harz*, D–2349 *Rhön*, D–2352 *Eifel* and D–2359 *Hunsrück*.

The M 20s entered service in 1929 on

Lufthansa's Basle-Mannheim-Frankfurt-Cologne-Essen-Amsterdam and Stuttgart-Geneva-Marseilles-Barcelona routes, two aircraft still being recorded as operational as late as 1942. From the middle 1930s the aircraft were switched to *Lufthansa*'s holiday services to the German Friesian Islands, operating the Frankfurt-Hanover-Bremen - Wilhelmshaven - Wangerooge - Norderney and Berlin-Kiel-Wyk-Westerland routes. Other less important services taken over by the M 20s were those from Hamburg to Kiel and Flensburg, and from Stuttgart to Freiburg.

During 1932 several M 20s were fitted with 640 hp BMW VIu engines and given the designation M 20b–2. Following the introduction of the new civil aircraft designation system in 1933, most of the M 20s received new registrations. D–1480 became D–UFON and later D–OAQE, D–1676 became D–UDAL and D–OFYZ, D–2205 became D–UNAH, D–2025 became D–UKUM, D–2026 became D–UREK, D–2285 became D–UVOX and D–ORAD, D–2290 became D–UXYN, D–2341 became D–UKIP, D–2349 became D–UMOK and D–OFOT, D–2352 became D–UJAR, and D–2359 became D–UHEN and D–OCIR. In 1935 D–UKUM *Westerwald* was bought by *Hansa Flugdienst*, and in 1937 D–UKIP became PP–VAK *Acegua* after being sold to the Brazilian airline, *Varig*.

Specification (M 20b)

SPAN	25·50 m	(83 ft 8 ins)
LENGTH	15·90 m	(52 ft 2 ins)
HEIGHT	4·20 m	(13 ft 9¼ ins)
EMPTY WEIGHT	2,700 kg	(5,952 lbs)
LOADED WEIGHT	4,600 kg	(10,141 lbs)
MAXIMUM SPEED	175 km/h	(109 mph)
NORMAL RANGE	880 km	(547 miles)

Top left: *The tiny Messerschmitt M 17 (D–779) could easily be accommodated under the wing of the M 20a airliner.*

Left: *One of the Messerschmitt M 20b production aircraft used by* Lufthansa *on their Basle to Amsterdam and Stuttgart to Barcelona routes.*

M 21

DURING the negotiations that accompanied the amalgamation of BFW and Messerschmitt in 1927, the German Ministry of Transport attempted to influence Messerschmitt into producing a number of military aircraft. The aviation department of the *Reichswehrministerium* (Ministry of Defence) had experienced difficulty with other companies and it was envisaged that

The first of only two biplanes built by Messerschmitt, the M 21 two-seat trainer was not successful.

the new combine might provide them with a new source which would be willing to produce the aircraft they required.

Co-operation between the *Reichswehrministerium* and Heinkel was quite good, but relations between them and Dornier, Rohrbach, and particularly Junkers, were far from happy. An example of the latter was when the chief of the aviation department, Oberst Felmy requested the design of a new fighter. Sachsenberg, a Junkers director calmly told him; 'You need no fighter. Just take our G 24 and arm it with a dozen machine guns and you can use that as a fighter!'

Negotiations between the Transport Ministry and Messerschmitt resulted in two designs, the M 21 two-seat trainer and the M 22 bomber. The M 21 was in fact the first of only two biplanes ever produced by the Messerschmitt company. A neat single-bay design powered by either a 96 hp Siemens Sh 11 or 125 hp Sh 12 radial, the M 21 was not accepted for production.

Specification (M 21 with Sh 12)

SPAN	10·00 m	(32 ft 9¾ ins)
LENGTH	7·30 m	(23 ft 11½ ins)
HEIGHT	2·80 m	9 ft 2¼ ins)
EMPTY WEIGHT	480 kg	(1,058 lbs)
LOADED WEIGHT	740 kg	(1,632 lbs)
MAXIMUM SPEED	156 km/h	(97 mph)
NORMAL RANGE	600 km	(373 miles)

M 22

THE M 22 was a twin-engined medium bomber, the second of only two biplanes ever built by Messerschmitt. Rugged in appearance, the M 22 could carry a crew of three including a gunner in the nose and dorsal positions. The machine was powered by two 500 hp Bristol Jupiter radial engines built under licence by the Siemens company. The very antipathy of what Messerschmitt was trying to attain with his lightweight monoplanes, it is not surprising that the M 22 failed to gain a production contract, and only one was built.

Specification

SPAN	17·00 m	(55 ft 9¼ ins)
LENGTH	13·60 m	(44 ft 7½ ins)
HEIGHT	4·80 m	(15 ft 9 ins)
EMPTY WEIGHT	2,900 kg	(6,395 lbs)
LOADED WEIGHT	3,800 kg	(8,380 lbs)
MAXIMUM SPEED	220 km/h	(137 mph)
NORMAL RANGE	500 km	(311 miles)

Right: *A rare photograph of the Messerschmitt M 22 medium bomber powered by two Siemens Jupiter radials.*

M 23

As has already been mentioned in the introduction, the Bayerische Flugzeugwerke AG became bankrupt in June 1931. A small group of personnel from the Messerschmitt Flugzeugbau continued design and construction work, and it was during this period that development of the excellent M 23 two-seat sporting aircraft was continued.

During 1929 the *Deutsche Luftfahrtverband* (German Aero Club) issued new specification conditions for lightplanes to take part in the *Ostpreussenflug* (Circuit of East Prussia). Messerschmitt developed a new and improved version of the M 19 to meet these conditions under the designation M 23. The new project was basically a two-seat development of the M 19 with shorter and wider fin and rudder surfaces, and wings which could be folded back alongside the fuselage to facilitate easy storage.

Three major sub-types of the aircraft were produced, the first being the M 23a which was powered either by a 38 hp ABC Scorpion two cylinder engine or a 45 hp Salmson AD 9 nine cylinder radial. The Messerschmitt M 23b was an improved version with curved upper fuselage decking. A large number of engines were fitted to the aircraft including the 82 hp Siemens Sh 13, the 86 hp Armstrong Siddeley

Top: *An early production Messerschmitt M 23a powered by the two cylinder ABC Scorpion engine.*

Right: *At least one Siemens Sh 13 powered M 23b was fitted with floats.*

Genet, the 95 hp ADC Cirrus III, the 115 hp ADC Cirrus Hermes, the 115 hp Siemens Sh 14 and the 50 hp BMW X.

Theo Cronweiss won the 1929 *Ostpreussenflug* race flying a Genet-powered M 23b and during April of that year Swiss Air Force pilots successfully flew the design like a fighter aircraft. Apart from being an excellent sporting monoplane, the M 23 also possessed fine aerobatic qualities. Willi Stör, who usually flew a BFW U 12, entered his new M 23b for the 1929 *Deutsche Kunstflugmeisterschaft* (German Aerobatic Competition) and became second behind Gerhard Fieseler who was flying his own F.1 Tigerschwalbe design. Stör later became a popular figure at other air shows with his M 23b.

Perhaps the greatest successes gained by the M 23 were the winning of the 1929 and 1930 Circuit of Europe races. Four aircraft were entered in the earlier competition. These were piloted by Fritz Morzik (racing number A4), von Dungern (A3), Offermann (A2) and Theo Cronweiss (A5). The first three were powered by Siemens Sh 13 five cylinder radials, the latter by an Armstrong Siddeley Genet of similar configuration. The race was won by Morzik, the other pilots coming respectively sixth, eighteenth and twenty-ninth. Fritz Morzik later reached the rank of *Generalmajor* and commanded the *Luftwaffe*'s transport forces during the Demyansk and Stalingrad operations.

The successful conclusion of this race led Messerschmitt to develop an improved variant for the 1930 competition. Designated M 23c, the new machine differed in having an enclosed cockpit and could be powered by a 110 hp Argus As 8 four

Above: *A Cirrus III powered M 23b of the Bismarck Flying School at Schleissheim. This aircraft crashed on 26 July 1929 at Augsburg.*

Below: *The Messerschmitt M 23c could easily be distinguished by its enclosed cabin. The machine was powered by an Argus As 10 engine.*

Offermann's M 23c which competed in the 1930 Circuit of Europe race. The aircraft was registered D–1886 and carried the racing number C5.

cylinder in-line engine or a 150 hp Siemens Sh 14a seven cylinder radial.

Seven M 23cs were entered for the 1930 Circuit of Europe race which began on 29 September. The pilots, racing numbers and registrations were: Fritz Morzik (B3—D-1883), von Freyberg (C3—D-1884), von Waldau (C6—D-1885), Offermann (C5—D-1886), Krüger (E8—D-1887), von Massenbach (C7—D-1888) and Polte (F2—D-1889). Apart from the M 23s, Germany also entered an Albatros L.100 and L.101, a Darmstadt D-18, an Arado L.2a, a Junkers A 50 and a Klemm L.25E. Apart from the German entries, eight British, five French, twelve Polish and one Swiss, Belgian, Spanish and American aircraft were also entered for the contest. Again Fritz Morzik beat all-comers and during the technical examinations, achieved the excellent average of 427 points out of a possible 500.

The M 23's reputation was further enhanced by the flights of Ernst Udet with the film stars Luis Trenker and Leni Riefenstahl to Greenland, and solo to Africa. Another very important trip was that undertaken by the Messerschmitt pilot, Erwin Aichele and his wife. They flew a distance of 8,000 miles around the Mediterranean in a M 23c during 1933 without experiencing any mechanical difficulties.

Apart from its use as a racing and aerobatic aircraft, the M 23 was also sold to several private pilots, perhaps the most important of these being Rudolf Hess. It was Messerschmitt that later made a Bf 110 available to Hess to enable him to fly to England on 10 May 1941 in an attempt to end the Second World War.

Specification

SUB-TYPE	LENGTH	EMPT. WT	LOAD. WT	MAX SP.	N. RANGE
M 23a Scorpion	6·65 m 21′ 9¾″	220 kg 485 lbs	440 kg 970 lbs	130 km/h 81 mph	920 km 572 miles
M 23a Salmson	6·65 m 21′ 9¾″	250 kg 551 lbs	480 kg 1058 lbs	140 km/h 87 mph	810 km 503 miles
M 23b Sh 13	6·50 m 21′ 4″	325 kg 717 lbs	595 kg 1312 lbs	162 km/h 100 mph	800 km 497 miles
M 23b Genet	6·66 m 21′ 10¼″	285 kg 628 lbs	535 kg 1180 lbs	165 km/h 102 mph	800 km 497 miles
M 23b Cirrus	6·43 m 21′ 1¼″	355 kg 783 lbs	600 kg 1323 lbs	180 km/h 112 mph	700 km 435 miles
M 23b Hermes	6·43 m 21′ 1¼″	360 kg 794 lbs	600 kg 1323 lbs	185 km/h 115 mph	700 km 435 miles
M 23b Sh 14	6·35 m 20′ 10″	370 kg 816 lbs	670 kg 1477 lbs	185 km/h 115 mph	1000 km 621 miles
M 23b BMW X	6·50 m 21′ 4″	308 kg 679 lbs	600 kg 1323 lbs	140 km/h 87 mph	— —
M 23c As 8	7·00 m 22′ 11¼″	320 kg 706 lbs	600 kg 1323 lbs	175 km/h 109 mph	750 km 466 miles
M 23c Sh 14a	7·00 m 22′ 11¼″	400 kg 882 lbs	600 kg 1323 lbs	220 km/h 137 mph	— —

SPAN (M 23a and b)	11·80 m	(38 ft 8½ ins)
SPAN (M 23c)	12·00 m	(39 ft 4½ ins)
HEIGHT (M 23a and b)	2·30 m	(7 ft 6½ ins)
HEIGHT (M 23c)	2·40 m	(7 ft 10½ ins)

M 24

THE MESSERSCHMITT M 24 was a development of the M 20 airliner built during 1929. Although considerably smaller than the M 20, the M 24 was designed to carry eight passengers. It was basically similar to the earlier machine in appearance but could be easily distinguished by the reduced number of cabin windows.

Two variants were proposed, the M 24a which could be powered by a 320 hp BMW Va or a 280/310 hp Junkers Jumo L 5 liquid cooled in-line engine, and the M 24b which employed a 525 hp BMW Hornet or 500 hp Siemens Jupiter radial. Only two Messerschmitt M 24as were built, these being D–1767 (w/nr 445) powered by a Junkers L 5 engine and D–1853 (w/nr 446) with a BMW Va. The former aircraft was delivered to Nordbayerische Verkehrsflug in 1930 and operated the Dresden-Chemnitz-Plauen-Nuremberg service for three years before being destroyed in 1934. The other machine was delivered to the Deutsche Versuchsanstalt für Luftfahrt and used for test purposes.

It is not known how many of the faster and heavier M 24b variant were completed, although it was probably either two or three. Two aircraft were definitely fitted with BMW built Pratt and Whitney Hornet engines, one of these having twin floats. The Siemens Jupiter engined version may never have been completed.

Specification (M 24a)

SPAN	20·60 m	(67 ft 7 ins)
LENGTH	12·80 m	(42 ft 0 ins)
EMPTY WEIGHT	1,600 kg	(3,527 lbs)
LOADED WEIGHT	2,800 kg	(6,172 lbs)
MAXIMUM SPEED	195 km/h	(121 mph)
NORMAL RANGE	900 km	(559 miles)

The Junkers L 5 powered M 24a before it was registered D–1767 and delivered to Nordbayerische Verkehrsflug in 1930.

M 26

THE EXCELLENT but unsuccessful M 24 was followed by the M 25, a project for a lightplane to be built to the requirements of Ernst Udet. This was never completed, but the M 26 was a small passenger carrying aircraft based on the M 18 and M 24. A high-wing cabin monoplane with provision for two or three passengers, the M 26 was powered by a single 100 hp Siemens Sh 11 seven cylinder radial engine. As far as is known, only one aircraft was built, this making its first flight in 1930.

It is probable that the only reason for the failure of the M 26 was the financial crisis that was then assailing Bayerische Flugzeugwerke. An improved version of the design was also projected, employing all-metal construction and designated Messerschmitt M 30. This machine was to be powered by a 175 hp Wright Whirlwind radial, but no production was undertaken.

Specification

SPAN	12·40 m	(40 ft $8\frac{1}{4}$ ins)
LENGTH	7·15 m	(23 ft $5\frac{1}{2}$ ins)
HEIGHT	2·30 m	(7 ft $6\frac{1}{2}$ ins)
EMPTY WEIGHT	460 kg	(1,014 lbs)
LOADED WEIGHT	700 kg	(1,543 lbs)
PERFORMANCE	*No details available*	

Top: *The BMW Hornet powered M 24b. It is not known how many M 24bs were built although it is thought to have been two or three.*

Left: *Only one Messerschmitt M 26 was built, the neat high-wing monoplane being powered by a Siemens Sh 11 radial.*

Facing page: *A photograph of the M 27b in flight. This aircraft carries the registration D–1979.*

M 27

Following the success of the M 23 lightplane, Messerschmitt developed a further improved version with provision for increased load carrying capacity. The M 27 was basically similar to the M 23b apart from new and more rounded fin and rudder surfaces and a modified, spatted undercarriage. Two variants were built; the M 27a powered by a 110 hp Siemens Sh 12 radial and the M 27b which was fitted with a 120 hp Argus As 8 engine.

In 1923 the M 27 won the Deutschland Competition and in the following year came second in the Circuit of the Zugspitz mountain race. Despite these successes, the M 27 was not built in quantity.

Specification (M 27b)

SPAN	12·00 m	(39 ft 8½ ins)
LENGTH	7·90 m	(25 ft 11 ins)
HEIGHT	2·40 m	(7 ft 10½ ins)
EMPTY WEIGHT	420 kg	(926 lbs)
LOADED WEIGHT	720 kg	(1,588 lbs)
MAXIMUM SPEED	200 km/h	(124 mph)
NORMAL RANGE	700 km	(435 miles)

M 28

The Messerschmitt M 28 was a rather obscure single-engined mailplane designed for use by *Lufthansa*. It was an elegant low-wing monoplane with a fixed spatted undercarriage. Metal construction was used throughout apart from the control surfaces and the trailing edge of the wings aft of the main spar which were fabric covered. The aircraft was powered by an uncowled 525 hp Pratt and Whitney Hornet nine cylinder radial engine built under licence by the BMW company.

As far as is known, only two aircraft appear to have been built, both during 1930. Of these, only one aircraft has been identified, this being D–2059 (w/nr 527) which was delivered to *Lufthansa* during 1932. The aircraft was used on the airline's internal mail services until 1935 when they were returned to the *Reichsluftfahrtministerium* (German Aviation Ministry). When first delivered the Messerschmitt M 28 shared with the Focke-Wulf A 43 Falke the honour of being the fastest German civil aircraft.

Specification

SPAN	15·50 m	(50 ft 10¼ ins)
LENGTH	10·00 m	(32 ft 9½ ins)
EMPTY WEIGHT	1,160 kg	(2,557 lbs)
LOADED WEIGHT	2,750 kg	(6,062 lbs)
MAXIMUM SPEED	255 km/h	(158 mph)
NORMAL RANGE	1,900 km	(1,180 miles)

This Messerschmitt M 28 high-speed mailplane was delivered to Lufthansa *in 1932.*

M 29

Following his success in the first and second Circuit of Europe races in 1929 and 1930, Messerschmitt decided to develop a new aircraft to compete in the third contest to be held in 1932. The new machine, which was designated M 29, was an attractive low-wing monoplane with a fixed spatted undercarriage and an enclosed cockpit for the crew of two. Two variants were built, the M 29a powered by a 150 hp Siemens Sh 14a radial and the M 29b which was fitted with a 150 hp Argus As 8R engine.

The German team for the third *Challenge de Tourism Internationale* as it was officially known, comprised six Heinkel He 64s and four M 29s. The aircraft were all painted bright scarlet with white registrations and red and white striped fin and tailplane surfaces. The He 64s were registered D–2260 and from D–2301 to D–2305 and the M 29s were D–2306 (racing number C3), D–2307 (B2), D–2308 (B3) and D–2309 (B4). Four pilots were nominated to fly the M 29s; Fritz Morzik, Rheinhold Poss, Kamerad Kreuzkamp and Oskar Dinort.

The four M 29s were delivered eight days before the beginning of the race which was scheduled for 27 August 1932. After test flying the machine, Morzik was very impressed, although he considered that the horizontal stabiliser, the whole of which moved to provide control, might

Above: *A Messerschmitt M 29a with a M 29b in the background. Both machines were painted bright red with white stripes on the tailplane and rudder surfaces.*

Below: *The M 29b differed from the M 29a in being powered by an Argus As 8R engine.*

give trouble. Apart from this feature, Morzik found the M 29 was light and sensitive on the controls and possessed an exceptional performance.

Shortly afterwards, Morzik flew his aircraft (D–2307) to Brunswick, but when he arrived he learned that Kreuzkamp's machine (D–2308) had disintegrated in the air for no apparent reason, killing the pilot. On the following day, Poss's aircraft broke up as it came in to land at Schleischeim, but although the observer, Starchinsky was killed, Poss managed to escape by parachute. After recovering, Poss was able to explain that the disintegration had begun in the tailplane, eventually spreading to the remainder of the aircraft.

Following these accidents, the other two M 29s were grounded pending the strengthening of their tail surfaces. Morzik switched to a He 64c for the competition which was eventually won by Hans Seidemann flying a similar aircraft (D–2260, racing number A9) in the record time of three days. After modification, the other two M 29s flew quite successfully, although the precise cause of the destruction of the aircraft has never been revealed. At least one other aircraft was built, a M 29b (D–2259).

Specification (M 29b)

SPAN	11·00 m	(36 ft 1 ins)
LENGTH	7·75 m	(25 ft 5 ins)
HEIGHT	2·00 m	(6 ft 6¾ ins)
EMPTY WEIGHT	390 kg	(860 lbs)
LOADED WEIGHT	700 kg	(1,544 lbs)
MAXIMUM SPEED	260 km/h	(162 mph)
NORMAL RANGE	700 km	(435 miles)

M 31

ALTHOUGH Messerschmitt developed a number of beautiful lightplanes, perhaps the most attractive was the M 31. This was a further development of the M 19, M 23, M 27 series, possessing even sleeker lines. The tandem two-seat design was powered by a single 60 hp BMW Xa five cylinder radial engine. It was, like its predecessors, a neat low-wing monoplane with a braced tailplane and fixed undercarriage.

Only one prototype of the M 31 was built, this being exhibited at the German Aerosport exhibition held in 1933. Despite its undoubted sleekness of line, no production orders were forthcoming.

The Messerschmitt M 32 was a project for a tandem two-seat biplane trainer developed at the instigation of the *Verkehrsministerium*. Mainly because of Messerschmitt's disinterest with biplanes, the project was never built.

Specification (M 31)

SPAN	12·00 m	(39 ft $4\frac{1}{2}$ ins)
LENGTH	7·85 m	(25 ft 9 ins)
HEIGHT	1·90 m	(6 ft $2\frac{3}{4}$ ins)
EMPTY WEIGHT	335 kg	(739 lbs)
LOADED WEIGHT	650 kg	(1,433 lbs)
MAXIMUM SPEED	170 km/h	(106 mph)
NORMAL RANGE	700 km	(435 miles)

Right: *Perhaps the most attractive of the many Messerschmitt lightplanes was the tiny M 31 powered by a BMW Xa radial.*

M 33

THE STORY of the 'People's Car' designed by Dr Ferdinand Porsche is now well known. Projected well before the Second World War, the Volkswagen is still going strong, albeit in a slightly modified form.

What is perhaps not so well known is that Messerschmitt attempted to build a 'People's Aircraft' to be known as the M 33. This was basically a wing, beneath which was mounted a gondola containing the pilot and a 15 hp DKW engine. This was to drive a two-bladed airscrew mounted just above the centre of the wing leading edge via a special gearing device. The tail was mounted on a steel tube which was attached to the wing trailing edge. A mock-up of the design was completed in 1933, but no production was undertaken.

Specification

SPAN	9·26 m	(30 ft $4\frac{1}{2}$ ins)
LENGTH	5·30 m	(17 ft $4\frac{3}{4}$ ins)
HEIGHT	1·90 m	(6 ft $2\frac{3}{4}$ ins)

(*No details available on weights and performance*)

M 35

THE LAST of a series of excellent lightplanes designed by the Messerschmitt company, the M 35 was a further development of the M 23, M 27, M 31 series. The M 35 was very similar in appearance to the M 27, but could be distinguished by its single leg spatted wheels in place of the extensively braced units of the latter. Single and twin seat variants of the M 35 were built, the machine subsequently achieving fame for its advanced aerobatic displays.

Two variants of the aircraft were completed; the M 35a which was powered by a 150 hp Siemens Sh 14a seven cylinder radial and the M 35b which employed a 135 hp Argus As 8b four cylinder in-line engine. The M 35 was also projected with a 225 hp Argus As 17 power unit. The first of 15 Messerschmitt M 35s were completed during the summer of 1933, the aircraft first being revealed to the public at the *Aerosalon* held in Geneva in 1934.

The first success achieved by the M 35 came in 1934 when the *Nazi* leader Rudolf Hess won the Zugspitz Trophy. On 11 and 12 May 1935 the Messerschmitt pilot Willi Stör won the *Deutsche Kunstflug-meisterschaft* (German Aerobatic Championship) held in Stuttgart flying a M 35a. Stör entered his M 35a (D–EQAN) for the next year's contest which was held in Munich on 11 and 12 July. The machine

Top: *Only a mock-up of the strange-looking M 33 'People's Aircraft' was completed.*

Right: *The M 35, a more powerful version of the M 31, was one of the finest aerobatic machines of the pre-war era.*

was painted bright yellow with blue fanwise stripes on the upper wing surfaces. The BFW badge (a symbolic bird) and the Olympic Games rings were painted on the forward fuselage, the latter commemorating the 1936 Games being held in Berlin.

Again Willi Stör won the contest with 120·16 points, Gerd Achgelis flying a Fw 56 Stösser coming second with 108·66 and Graf von Hagenburg in a Fw 44 Steiglitz third with 107·80 points. The women's competition was won by Vera von Bissing, also flying an M 35. Success followed success, the M 35 achieving prominence during the Zurich flying meeting of 1937.

The M 35 was later flown very successfully by Swiss, Italian and French pilots, and until the beginning of the Second World War remained one of the finest aerobatic aircraft in the world. The aerobatic qualities of the Bf 109 were to owe much to the success of the M 35.

Specification (M 35a)

SPAN	11·57 m	(37 ft 11½ ins)
LENGTH	7·48 m	(24 ft 6 ins)
HEIGHT	2·75 m	(9 ft 0¼ ins)
EMPTY WEIGHT	500 kg	(1,102 lbs)
LOADED WEIGHT	800 kg	(1,764 lbs)
MAXIMUM SPEED	230 km/h	(143 mph)
NORMAL RANGE	700 km	(435 miles)

M 36

FOR A CONSIDERABLE TIME before Erhard Milch's appointment as Secretary of State for Air in April 1933, there had existed a long-standing feud between him and Willi Messerschmitt. After Milch's appointment, it became obvious that the Bayerische Flugzeugwerke AG was unlikely to receive any contracts for the development of new aircraft, and therefore the company attempted to seek orders from abroad.

In June 1933 Messerschmitt's co-director, Kokothaki, travelled to Bucharest and gained an order from the Rumanian ICAR company for the design of a small transport

aircraft. The project, which was conceived around a 380 hp Gnome-Rhone Jupiter radial, was designated M 36 by Messerschmitt. It was a neat high wing monoplane with a fixed spatted undercarriage of Messier design attached to the wing.

Towards the end of 1933 the *Intreprindere Constructii Aeronautice Romane* (ICAR) company bought the drawings of the aircraft outright and built at least one example under the designation IAR 36. The wings were constructed of Rumanian pine with plywood covering and the fuselage and tail surfaces were built of welded steel tube with metal covering forward of the passenger cabin and fabric aft.

After the completion of trials during the winter of 1934–35 the aircraft was delivered to the Rumanian airline, LARES as YR–ACS. It was powered by a 380 hp Armstrong Siddeley Serval radial and could carry a crew of two and six passengers. The machine operated several internal services including that between Bucharest and Cernauti in northern Rumania.

The fact that BFW had gained a foreign development contract for the M 36 led to bitter complaints being aimed at the company by Major Wimmer of the RLM's Technical Office. Messerschmitt replied to this criticism by saying that since no German originated contracts had been forthcoming, he was forced to turn to foreign countries for orders to keep his design department in work. As a result the BFW company were asked to participate in a design competition for a single-engined fighter which eventually emerged as the Bf 109.

Specification (M 36)

SPAN	15·40 m	(50 ft 6¼ ins)
LENGTH	9·82 m	(32 ft 2½ ins)
HEIGHT	3·50 m	(11 ft 9½ ins)
EMPTY WEIGHT	1,350 kg	(2,976 lbs)
LOADED WEIGHT	2,250 kg	(4,960 lbs)
MAXIMUM SPEED	240 km/h	(149 mph)
NORMAL RANGE	700 km	(434 miles)

Bf 108 Taifun

THE Bf 108, which was designed for the 1934 *Challenge de Tourism Internationale* race was the first Messerschmitt aircraft to be built of all-metal construction. Originally known as the Messerschmitt M 37, the machine was later allocated the type number 8–108 under the RLM's new comprehensive designation system.

In June 1934 the prototype Bf 108A (D–ILIT) made its initial flight. It was a beautiful low-wing monoplane owing much to the earlier Messerschmitt sporting aircraft, but introducing several advanced features including an enclosed cabin for

Right: *The first prototype of the excellent* Taifun *communications aircraft was D–ILIT, forerunner of the Bf 108A series.*

Left: *Only one example of the IAR (Messerschmitt) 36 is known. Carrying the registration YR–ACS, the machine was delivered to the Rumanian Air Line, LARES.*

the crew of four and an outwards retracting undercarriage. A total of seven Bf 108As were completed, six of them (registered D–ILIT, D–IBRE, D–IBUM, D–IMUT, D–IJES and D–IZAN) being powered by a 225 hp Hirth HM 8U engine driving a fixed pitch three-bladed airscrew. The other aircraft (D–IGAK) was powered by a 220 hp Argus As 17B engine and early in 1935 D–IZAN was fitted with a 160 hp Siemens Sh 14A radial.

The appearance of the sleek new Messerschmitt aircraft caused a sensation with other designers and the aviation minded public alike. It proved an extremely pleasant machine to fly and, with its Handley Page leading edge slots, possessed an extremely wide speed range spectrum, ranging from 38 to 181 mph.

All seven Bf 108As were delivered to the German team for the fourth *Challenge de Tourism Internationale*, but after one aircraft crashed, the team manager, Theo Osterkamp, attempted to have the design withdrawn from the contest. His decision was overruled by *Major* Loeb, but because of its heavier all-metal construction, the Bf 108 failed to beat the lighter wood and fabric designs which were favoured by the rules of the contest.

In 1935 Elly Beinhorn, the famous German woman pilot, flew a Bf 108A (D–IJES) from Berlin to Constantinople

Top left: *The prototype Bf 108B (D–IELE) was fitted with a 160 hp Siemens Sh 14A radial engine.*

Left: *A standard production Bf 108B in pre-war markings outside a Messerschmitt hangar.*

Facing page: *The* Geschwader Stab *of JG 2* Richthofen *used this Bf 108B for communications duties during 1940.*

and back in one day. Her aircraft was named *Taifun* (Typhoon) and this title was subsequently adopted for all future Bf 108 production aircraft.

During the winter of 1935–36 the first Bf 108B–0 left BFW's factory at Augsburg. The new variant introduced a tailwheel in place of the skid, revised fin and rudder contours, deletion of the upper tailplane bracing and dynamic balancing of the rear control surfaces. The prototype Bf 108B (D–IELE) was powered by a 160 hp Siemens Sh 14A radial engine in an NACA cowling, but this proved unsuitable and was replaced by the 240 hp Argus As 10C or 270 hp As 10E unit in all subsequent aircraft. Early Bf 108B production aircraft featured a two-bladed fixed pitch airscrew but this was later replaced by a variable pitch unit of Argus design.

The Bf 108 really achieved prominence during the 1936 Olympic Games held in Berlin, competing in the various aviation rallies held during the Games. One machine, piloted by the Messerschmitt pilot, Brindlinger, was used to ferry film of the Olympic events to the Swedish press. Eventually the aircraft clocked-up 13,500 miles of trouble free flying.

During 1937, Elly Beinhorn flew a Bf 108 from Berlin to Capetown and back and in the International Circuit of the Oases race, *Taifuns* came second and fourth. In 1938 *Flugkapitän* Hans Seidemann successfully flew the Bf 108 in the Isle of Man race and many other firsts were gained by the aircraft during the year. These included the *Hoggar Internationale Sternflug*, the Belgian International Königin-Astrid-Rennen contest in July and the International Flight Competition at Dinard in France a month later.

The success of the Taifun prompted the RLM to consider the mass production of the aircraft for use by the *Luftwaffe* in the fast transport and communications rôles. A large number of Bf 108s were later used by many operational units. Most *Gesch-*

wader Stabs flights had at least one aircraft on hand, and the *Luftdienstverbänd*, which undertook a wide variety of duties including target towing, rescue and supply operations, used the machine in considerable numbers.

In 1941, Messerschmitt proposed a special record-breaking version of the Taifun, to be known as the Bf 108C. The aircraft was to be powered by the experimental 400 hp Hirth HM 512A engine driving an Argus variable pitch propeller. It is doubtful whether any Bf 108C was completed.

Following the invasion of France and the capture of many aircraft plants, it was decided to transfer production of several strategically unimportant aircraft to such factories. Amongst these machines was the Bf 108, production of which was transferred to the S.N.C.A. du Nord factory at Les Mureaux. The French-built variant was probably designated Bf 108D, and before the end of the war a total of 885 Bf 108s had been completed.

In addition to being delivered to the *Luftwaffe*, about 50 Bf 108s were exported to foreign countries including Bulgaria, Hungary, Japan, Rumania, Switzerland, Russia and Yugoslavia. Two German Embassy aircraft in Britain (D–IDBT and D–IJHW) were impressed into the RAF in September 1939 as ES955 and AW167, and a third aircraft (G–AFRN) became DK280. Fifteen Bf 108s were also captured by the RAF after the war and allocated the serial numbers VM495, VM502, VM508 and VM851 to VM862.

Production of the Bf 108 continued in France after the war, 285 machines being built for the *Armée de l'Air* and the *Aéronavale*. In France, the Bf 108 was re-designated as the Nord 1000 and variants re-engined with the 240 hp Renault 6Q–10 and 6Q–11 were known as the Nord 1001 Pingouin I and Nord 1002 Pingouin II respectively. Two other experimental variants were proposed, the Nord 1003 with a 240 hp Potez 6D engine and the Nord 1004 with a 305 hp Potez unit.

Specification (Bf 108B)

SPAN	10·62 m	(34 ft 10¼ ins)
LENGTH	8·29 m	(27 ft 2½ ins)
EMPTY WEIGHT	880 kg	(1,941 lbs)
LOADED WEIGHT	1,352 kg	(2,981 lbs)
MAXIMUM SPEED	303 km/h	(188 mph)
NORMAL RANGE	1,000 km	(621 miles)

Bf 109B and C

The Bf 109 V4 (D–IOQY) was the first prototype to carry three MG 17 machine guns.

To THE LAYMAN the name of Messerschmitt must inevitably conjure forth a vision of a small single-engined fighter, which became the arch enemy of the Spitfire during the Battle of Britain. Known officially as the Bf 109, this fighter was remarkable in that it was the first military machine to be built by the Messerschmitt company. Apart from its justifiable fame, the Bf 109 was also built in larger numbers than any other aircraft with the possible exception of the Russian Ilyushin Il–2.

As has been described in the M 36 history, the Messerschmitt company had been forced to solicit orders from abroad in order to keep their design in work. In an attempt to prevent this re-occurring, the RLM issued their specification for a single-engined fighter monoplane to Messerschmitt late in 1933. As the company had no experience of military aircraft, and such famous names as Arado, Focke-Wulf and Heinkel were also asked to participate, it was widely felt that they would have no chance in the forthcoming competition.

It will however be remembered that Messerschmitt had considerable experience with the building of fast monoplane touring aircraft, and this they embodied into the Bf 109. The new aircraft was basically a neat low-wing monoplane designed around the most powerful German aero engine available at that time, the 610 hp Junkers Jumo 210A.

The prototype, the Bf 109 V1, D–IABI (w/nr 758), made its first flight in September 1935 with *Flugkapitän* Knötsch at the controls. Because of the shortage of Jumo 210 units, the Bf 109 V1 was powered by a 695 hp Rolls-Royce Kestrel V engine. In October the machine was flown to Rechlin, and then on to Travemünde where it was to compete with its rivals; the Ar 80 V1, the Fw 159 V1 and the He 112 V1.

Both the Ar 80 (which had a fixed undercarriage) and the Fw 159 (which featured a parasol wing) were quickly ruled out of the competition, but there remained little to choose between the other two. Eventually, and somewhat to the surprise of the pundits who favoured the He 112, an order was placed for ten prototypes of both aircraft.

The Bf 109 V2, D–IUDE (w/nr 809), first flew in January 1936, differing in being powered by a 610 hp Jumo 210A engine. The aircraft had provision for two 7·9 mm MG 17 machine guns above the engine cowling; the same arrangement as that proposed for the unbuilt Bf 109A production model. The V3 D–IHNY (w/nr 810) was generally similar, joining the test programme in June 1936. The Bf 109 V4, D–IOQY (w/nr 878), was fitted with a third MG 17 gun firing through the spinner, the intention being to replace this weapon with a 20 mm MG FF cannon at a later date.

During March 1936 the RLM placed a production order for the Bf 109, the first contract received by Messerschmitt for a

Left: *This photo is said to depict the experimental Bf 108C which was fitted with a 400 hp Hirth HM 512 engine.*

military aircraft. An armament of three MG 17 machine guns was retained by the Bf 109 V5, D–IEKS and V6, D–IHHB, the seventh aircraft D–IALY being a prototype for the Bf 109B production model. The first of a small batch of 109B–0 pre-production aircraft were delivered to the *Luftwaffe* early in 1937, being followed by the generally similar B–1. This machine was powered by a 635 hp Jumo 210D engine driving a fixed pitch airscrew and carried three MG 17 machine guns. The first major production model was the Bf 109B–2 which was powered by a 640 hp Jumo 210G engine which drove a two-blade variable pitch propeller.

The first deliveries to a *Luftwaffe* unit were made to I. *Gruppe*/JG 132 *Richthofen* based at Döberitz near Berlin. There was little doubt that this unit now possessed the most potent single-engined fighter in the world, and Goebbels's propaganda ministry was not slow to capitalise on this fact.

The Bf 109 V1 had made a brief appearance at the 1936 Olympic Games, but it was at the Zurich International Flying Meeting that the fighter made its first real impact. Five Messerschmitt aircraft took part in the meeting which was held between 23 July and 1 August 1937. These were the DB 600A powered Bf 109 V10, D–ISLU,

Top: *The Bf 109 V7 (D–IALY) was the first prototype for the proposed Bf 109B series.*

Left: *A standard production Bf 109B–1 in early* Luftwaffe *markings.*

and V13, D–IPKY, two Bf 109B–1s and a B–2. After Ernst Udet crashed the V10, *Major* Hans Seidemann, who, it will be remembered, achieved fame in an He 64 in 1932, won the Circuit of the Alps race flying a Bf 109B–2. The climb and dive competition was won by *Dipl Ing* Frank flying the V13, and Messerschmitt also came first in the speed competition and the team race.

In April 1937 the first Bf 109B–1 and B–2 fighters were delivered to 2.*Staffel* J/88 of the *Condor Legion* in Spain. The new fighter quickly gained air superiority over the Russian aircraft in use by the Republicans, and demonstrated for the first time in combat its formidable performance. In September, 1.J/88 re-equipped with the Bf 109, transferring its He 51s to the re-formed 4.*Staffel.*

The Spanish Civil War had a most important effect on subsequent *Luftwaffe* tactics, not least of which were the combat formations adopted by the *Jagdflieger.* Initially German fighter pilots flew in tight parade-ground formations, but this soon proved much too restrictive in combat. Remembering the lessons gained in air fighting during the First World War, a loose formation of four aircraft was developed in which each aircraft was placed like the fingertips of an outstretched hand. The fighters were spaced about 650 feet apart and controlled by FuG 7 radio equipment. The formation, known as the *Schwarm* within the *Luftwaffe* proved very successful, being copied by all major air forces.

A Staffel *of Bf 109B–2s from I./JG 132* Richthofen *at Carlsbad, Czechoslovakia, 1937.*

About this time, reports began to reach Germany of the RAF's new Hurricane fighter which carried an armament of eight machine guns. In an attempt to increase the armament of the Messerschmitt fighter, the Bf 109 V8 was fitted with four MG 17 guns, two of which were carried in the wing roots. The ninth prototype was similar, but carried two 20 mm MG FF cannon in the wings.

These two prototypes led to the Bf 109C–0 which differed from the B–2 merely in having the armament first carried by the V8. The Bf 109C–1 was similar, but the C–2 introduced a fifth MG 17 machine gun to fire through the spinner. The proposed Bf 109C–3 was to have carried the armament of the V9 and the unbuilt C–4 was to have been fitted with an engine-mounted MG FF cannon.

During the summer of 1939, all existing *Luftwaffe Gruppen* were redesignated to conform with the new *Luftflotte* system of operational commands. When war was declared in September 1939, several units were still equipped with the Bf 109C. These included I./JG 21 at Jesau, and I., II. and III./ZG 26 at Dortmund, Werl and Lippstadt respectively. ZG 26 *Horst Wessel* had been formed originally from JG 234.

Specification (Bf 109C–1)

SPAN	9·85 m	(32 ft 4¼ ins)
LENGTH	8·70 m	(28 ft 6½ ins)
HEIGHT	3·40 m	(11 ft 2 ins)
LOADED WEIGHT	2,300 kg	(5,071 lbs)
MAXIMUM SPEED	440 km/h	(273 mph)
SERVICE CEILING	10,000 m	(32,808 ft)

The Bf 109C–1 differed from the earlier B-series in carrying four MG 17 machine guns.

Bf 109D and E

Mention has been made previously of the Bf 109 V10 and V13 which had been experimentally fitted with 960 hp DB 600A engines. The V10 crashed, but the V13 was later fitted with the further improved DB 601A engine which, specially boosted to provide 1,650 hp for take-off, broke the world speed record for landplanes on 11 November 1937 by attaining a speed of 379·38 mph.

The Bf 109 V11 and V12 were generally similar to the V10, being fitted with production variants of the 960 hp Daimler-Benz DB 600 engine. These prototypes achieved a maximum speed of 323 mph and were so successful that a production development, designated Bf 109D was considered. Relatively few Bf 109D–0 and D–1s were completed before production switched to the further improved Bf 109E, the proposed Bf 109D–2 and D–3 being abandoned.

This variant, known throughout the *Luftwaffe* as the 'Emil' was powered by the much improved 1,100 hp DB 601A engine which featured direct fuel injection and enhanced supercharging capacity. The Bf 109 V14, D–IRTT, which flew during the summer of 1938, was the first machine to be fitted with the DB 601A engine, carrying an armament of two MG FF cannon in the wings and two MG 17 machine guns above the engine. The Bf 109 V15, D–IPHR, was similar but carried a single MG FF cannon firing through the spinner and no wing guns.

The Bf 109E–0 pre-production aircraft began to leave Messerschmitt's Regensburg assembly lines towards the end of 1938, this and the E–1 carrying an armament of four MG 17 machine guns, although the later aircraft switched to MG FF cannon in the wing positions. The Bf 109E–1 and E–1/B fighter-bomber were certainly comparable to any other fighter extant, attaining a maximum speed of 354 mph

The DB 600A powered Bf 109 V13 (D–IPKY) which won the climb and dive competition at the 1937 Zurich flying meeting.

and an initial climb rate of 3,100 ft/min.

JG 234 *Horst Wessel* became the first unit to receive the Bf 109E in December 1938 and by the following September no less than 15 *Gruppen* were equipped with the variant. These were I.(J)/LG 2, I./JG 3, I./JG 20, I. and II./JG 26, I./JG 51, I./JG 52, I. and II./JG 53, I./JG 70, I./JG 71, I./JG 76, I. and II./JG 77 and II./ZG 1. Five *Gruppen* retained the Bf 109D-1 including I./JG 2, I./JG 1, I./ZG 2, I./ZG 52 and II./ZG 76—the former unit retaining the variant until early 1940. The naval *Staffeln* 5. and 6./ *Trägergruppe* 186 were equipped with a mixture of Bf 109Bs and Es.

On 1 September 1939, German forces invaded Poland and two days later Britain and France declared war. Several fighter units took part in the battle over Poland, but as most of the Polish Air Force had been wiped out on the ground, they were limited mainly to ground strafing. In December Bf 109s met RAF bombers for the first time when aircraft from II./JG 77 under *Major* von Bülow and 10.(N)/JG 26 under *Oblt* Steinhoff successfully intercepted a formation of Wellingtons over the Heligoland Bight. Only one unit, II./JG 77 took part in the Norwegian campaign, the short range of the Bf 109 being somewhat inadequate for the long distances involved.

Not many Bf 109D-1s were completed before production was switched to the improved E-series.

Prior to the invasion of Norway, a rationalisation of the *Luftwaffe*'s fighter units took place. JG 2 *Richthofen*, JG 3 *Udet*, JG 26 *Schlageter*, JG 52 and JG 53 *Pikas* were all brought up to a strength of three *Gruppen*, JG 70, JG 71 and JG 72 being simultaneously disbanded. A second *Gruppe* was also added to JG 51 and three entirely new groups were formed, I./JG 27 on 1 October 1939, II./JG 27 on 1 January 1940 and I./JG 54 *Grünherz* a month later.

On 10 May 1940 the *Wehrmacht* launched its greatest invasion operation to date when it marched into Holland, Belgium and France. Although fighting bravely, the small Dutch army and air force was quickly overwhelmed, and Belgium capitulated soon afterwards. Only one French fighter, the Dewoitine D 520, was capable of meeting the Bf 109 on anything like equal terms, and the Hurricanes of the BEF were so harried on the ground that they failed to put up any effective opposition.

During the fighting, two young *Luftwaffe* officers began to achieve prominence: *Hptm* Werner Mölders, commander of III./JG 53 and *Hptm* Adolf Galland, leader of III./JG 26. Both had flown with J/88 in Spain and successfully added to

Facing page: *The Bf 109 V15 (D-IPHR) was the second variant to be powered by the DB 601A engine with fuel injection.*

their victory score during the French campaign. Before being shot down by French fighters on 5 June and taken prisoner of war, Mölders had gained 25 victories.

On 27 May 1940, over the beaches of Dunkirk, Messerschmitt 109s at last met their match in the shape of a sleek single-engined fighter with beautiful lines: the Spitfire. Thus began a rivalry which was to exist into the late 1940s when Czech built Bf 109s of the newly created Israeli state clashed with Spitfires of the Egyptian Air Force over Palestine. The few Spitfire squadrons operational in 1940 had been held in reserve in England, being thrown into the battle over Dunkirk, in an attempt to bring back the remnants of the BEF. In the Spitfire, the RAF possessed an aircraft that was the equal and in many ways the superior of the Bf 109. The stage was set for the 'Few' to face the 'Many'.

Towards the end of 1939, the Bf 109E–3 had entered service. The new variant was powered by a DB 601Aa engine and carried an armament of three 20 mm MG FF cannon and two 7·9 mm MG 17 machine guns. The Bf 109E–4 had the engine-mounted MG FF deleted and the E–5 and E–6 were reconnaissance fighters with armament reduced to two MG 17 guns. The former was powered by the 1,100 hp DB 601A engine, the latter having the more powerful DB 601N which produced 1,200 hp for take-off.

The Battle of Britain can be said to have commenced with the formation of a *Luftwaffe* anti-shipping group under *Oberst* Johannes Fink at the end of June 1940. Apart from Fink's own KG 2, the group comprised two *Stukagruppen* plus JG 51 under *Oberst* Theo Osterkamp to which III./JG 3 under *Hptm* Walter Kienitz was attached. The first clash between Spitfires and JG 51's Bf 109Es came on 8 July 1940 when two of the former and four of the latter were shot down.

Two days before this, a further rationalisation of the *Jagdverband* had taken place. I./JG 1 became III./JG 27, I./JG 20 became III./JG 51, I./JG 21 became III./JG 54 and I./JG 76 became II./JG 54. For the Battle of Britain proper, the following *Geschwader* were operational: JG 2 under *Oberst* von Bülow, JG 3 under *Oberstlt* Viek, JG 26 under *Maj* Handrik, JG 27 under *Oberst* Ibel, JG 51 under *Oberst* Osterkamp, JG 52 under *Maj* von Merhart, JG 53 under *Oberstlt* von Cramon-Taubadel, JG 54 under *Maj* Mettig and II./JG 77 under *Hptm* Hentschel, the latter in Norway. On 27 July 1940, *Maj* Mölders took over JG 51, the unit subsequently being known as *Jagdgeschwader Mölders*.

It was not until the early days of August 1940 that the *Luftwaffe* clashed with the RAF in a really major action. Previous to this, fighting had been on a relatively minor scale, one of the most important actions concerning the Bf 109 being fought on 28 July when eight aircraft from JG 27, 51 and 53 were lost. Amongst these was the Bf 109E–3 piloted by *Maj* Mölders which was severely damaged by a Spitfire flown by Sqdr Ldr Malan of 74 Squadron. The wounded Mölders managed to crash land his machine near Wissant.

The first major action of the Battle was on 8 August 1940 when the *Luftwaffe* made several heavily escorted dive bomber attacks on British convoys, the RAF putting up a large number of defensive sorties.

A pre-production Bf 109E–0 is wheeled out on to a snow-covered airfield.

A total of 17 Bf 109s, and six Bf 110s were lost for the destruction of 13 Hurricanes and seven Spitfires. Most heavily engaged were the Bf 109Es of JG 27, who lost 11 aircraft including that of the *Kommanduer* of II. *Gruppe*, *Maj* Werner Andres.

The next major action came three days later when 16 Bf 109s were destroyed plus 11 Bf 110s, for the loss of about 30 British fighters. Most heavily engaged were the Bf 109Es of JG 2, JG 27 and JG 54. Another large-scale action followed on 12 August, and on Eagle Day (13 August) 32 Bf 109s and Bf 110s were lost for the destruction of 20 Hurricanes and Spitfires. Two days later the Bf 110s suffered very heavy losses, and it became obvious that the Bf 109 would have to escort the twin-engined fighter if the latter was to survive at all.

On 16 August the *Luftwaffe* lost 21 Bf 109s, and an equal number were destroyed two days later. Six days later 21 Bf 109Es were again lost including eight from JG 51. On 26 August JG 3, 52 and 53 fared badly and two days later 23 Bf 109s were lost including six from JG 3. Sporadic heavy actions continued, the *Luftwaffe* losing 26 Bf 109s at the end of the month including six aircraft from JG 3, six from JG 26 and eight from JG 77.

German fighter losses continued to be heavy during the early days of September, 25 Bf 109s being lost on 2nd, 22 on 5th and 27 on the 6th, the latter including nine from JG 27. On the following day the

Luftwaffe switched its attentions to bombing London which swiftly relieved the hard-pressed fighters of the RAF. On 15 September 23 fighters from a total of 80 aircraft were lost by the *Luftwaffe* but the heaviest Bf 109 losses of the campaign came on the last day of the month when 34 aircraft were destroyed. This total included five from JG 2, seven from JG 26, eight from JG 27 and four each from JG 51, 52 and 53.

From the beginning of October 1940 operations by the *Luftwaffe* began to tail off, emphasis switching to night attacks by bombers on British cities. Several *Luftwaffe* pilots became famous during the conflict including *Maj* Walter Oesau, commander of III./JG 51, *Hptm* Günther Lützow, commander of I./JG 3, *Hptm* Hermann-Friedrich Joppien of I./JG 51 and *Maj* Helmut Wick, commander of JG 2. In what must be considered one of the most notable combat operations on 28 November 1940, Wick was shot down by Flt Lt 'Cocky' Dundas of 609 Squadron and killed. Immediately afterwards Dundas himself was shot down by another of JG 2's Bf 109s.

During the winter of 1940–41 several new variants of the Messerschmitt fighter were delivered. The Bf 109E–4/B was a fighter-bomber with provision for four 110lb or one 550lb bomb. The Bf 109E–4/N was powered by a 1,200 hp DB 601N

Top: *The Bf 109E–1/B fighter-bomber carried a 550 lb SC 250 bomb beneath the fuselage.*

Right: *A Bf 109E–3 in flight. This was one of the more important versions of Messerschmitt's superb fighter.*

engine, the E–5 and E–6 have already been described and the E–7 was similar to the E–4/N apart from the introduction of drop tank attachment points. The Bf 109E–7/U2 was a heavily armoured ground-attack aircraft, the E–7/Z carried GM–1 power boosting, the E–8 introduced the 1,300 hp DB 601E engine and the E–9 was a reconnaissance fighter.

On 14 April 1941, the first squadron of I./JG 27 arrived at Ain el Gazala airfield in North Africa. Before long the complete *Gruppe* was operational, being fresh from operations over Malta and against Yugoslavia earlier in the year. The introduction of the Bf 109E in North Africa quickly established air superiority for the *Luftwaffe*. I./JG 27 was commanded by *Hptm* Eduard Neumann and under him, in 3.*Staffel* was a young pilot who was to become one of the most famous in the *Luftwaffe*. His name was Hans-Joachim Marseille.

When Hitler launched Operation Barbarossa, the invasion of the Soviet Union on 22 June 1941, seven *Luftwaffe Gruppen* retained the Bf 109E. These were II. and III./JG 27 under *Maj* Wolfgang Schnellmann, II. and III./JG 52 under *Maj* Hans Trübenbach and I., II. and III./JG 77 under *Maj* Bernhard Woldenga. Initial operations by the Bf 109Es were made using $4\frac{1}{2}$-lb SD–2 anti-personnel bombs and it was during the first of these that Schnellmann was killed. The command of JG 27 then passed to Woldenga.

One final variant of the basic Emil air frame was produced designated Bf 109T. This aircraft was designed to operate from the aircraft carrier *Graf Zeppelin* which, in the event, was never completed. Ten Bf 109T–0s were built by the Fieseler company at Kassel, the machine introducing manually folding outer wing panels, increased wing span, an arrester hook and catapult spools. A total of 60 Bf 109T–1s were produced, this variant being powered by the 1,200 hp DB 601N engine. The Bf 109T–2 was a conversion of the T–1 which had the carrier equipment deleted.

Several Bf 109T–1s were used successfully by I./JG 77 operating from Norway, and after conversion to T–2 standard, they passed, in April 1943, to *Jagdstaffel Helgoland.* Based on the tiny island of Heligoland, the *Staffel* (later known as 11./JG 11) helped to combat the early USAAF raids against the north of Germany.

Specification (Bf 109E–3)

SPAN	9·85 m	(32 ft $4\frac{1}{4}$ ins)
LENGTH	8·63 m	(28 ft $3\frac{3}{4}$ ins)
HEIGHT	3·40 m	(11 ft 2 ins)
EMPTY WEIGHT	2,005 kg	(4,421 lbs)
LOADED WEIGHT	2,505 kg	(5,523 lbs)
MAXIMUM SPEED	570 km/h	(354 mph)
NORMAL RANGE	663 km	(412 miles)
SERVICE CEILING	11,000 m	(36,091 ft)

Right: *The Bf 109E–7/U2 heavily armoured ground attack aircraft in flight. The heavily framed cockpit structure easily distinguishes the late-production Bf 109E.*

Left: *A* Staffel *of Bf 109E–3 fighters on a* Luftwaffe *airfield during the summer of 1940.*

Top: *The Bf 109E airframe (w/nr 5604) which was fitted with the 1,300 hp DB 601E engine in a new and more symmetrical cowling.*

Left: *The Bf 109 V17 (D–IWKU) which crashed during trials. The machine was fitted with the new rounded wing tips later sported by the Bf 109F.*

Right: *Without doubt, the Bf 109F (here represented by a pre-production F–O) was the most attractive variant of the Messerschmitt fighter.*

Bf 109F and H

ALTHOUGH a considerable number of Messerschmitt 109 variants were built, *Luftwaffe* pilots are generally agreed that the Bf 109F was the finest. The success of the Bf 109E led her designers to produce a further advanced version making use of the more powerful 1,300 hp DB 601E engine.

During the spring of 1940, a standard Bf 109E airframe (w/nr 5604) was fitted with the new engine, the whole being enclosed by a new and much more symmetrical cowling. The machine flew successfully at Augsburg-Haunstetten on 10 July 1940, being followed by the Bf 109 V17 and V18, both of which were fitted with a new wing with rounded tips. The machines also introduced a smaller rudder and a fully retractable tailwheel. Towards the end of the year the first Bf 109F–0s were delivered to the *Luftwaffe* for evaluation. both this and the F–1 production model being powered by the 1,200 hp DB 601N engine.

In January 1941 the first production aircraft were delivered to the *Luftwaffe*, although some difficulties were experienced with the new, unbraced tail unit. Several aircraft crashed due to the fin and rudder breaking away from the fuselage, and the remaining machines were returned to the factory for structural modification.

The Bf 109F–1 carried an armament of a 20 mm MG FF cannon firing through the spinner and two 7·9 mm MG 17 machine guns above the engine cowling. The F–2 had the MG FF weapon replaced by the faster firing 15 mm MG 151 weapon, the F–2/Z carried GM–1 power boosting and the F–2/Trop was fitted with tropical filters. The Bf 109F–3 was powered by the 1,300 hp DB 601E engine and the F–4 introduced the heavier 20 mm MG 151/20 cannon in place of the faster-firing MG 151 gun.

III./JG 26 under *Maj* Gerhard Schöpfel was the first unit to receive the Bf 109F during May 1941. Although the new variant was generally accepted to be an improvement on the Emil, much controversy surrounded its armament. The Bf 109E–4 had carried two 20 mm cannon, but the F model had only one, and many pilots, including the *Kommodore* of JG 26, Adolf Galland, considered this entirely inadequate. On the other hand Mölders, who was an excellent shot, thought the centrally-mounted battery was a great advantage. Although the single cannon

Top left: *A standard production Bf 109F–1.*

Left: *The Bf 109F–4 was the first variant to carry the engine-mounted 20 mm MG 151/20 cannon.*

Top right: *A bomb being loaded aboard a Bf 109F–4/B of 10.(Jabo) Staffel/JG 26 during 1942.*

Right: *The film pack from a Bf 109F–5 photographic reconnaissance aircraft being handed to a* Luftwaffe *ground crew following a sortie.*

proved very effective in the hands of such a superb shot, it lost its effectiveness when flown by an inexperienced pilot, and eventually the Bf 109F–4/R1 was introduced which carried two MG 151/20 cannon in underwing gondolas.

Following the invasion of Russia in June 1941, only two fighter *Geschwader* remained on the western front, JG 2 under *Maj* Wilhelm Balthasar and JG 26 under *Oberstlt* Adolf Galland. Apart from I./JG 1 which had been re-formed in December 1940 for the defence of Germany and I./JG 27 which had been transferred to North Africa, all other fighter *Gruppen* were based in Russia. These included JG 3 under *Oberst* Günther Lützow, JG 27, JG 51 under *Oberst* Werner Mölders, JG 52, JG 53 under *Maj* Günther von Maltzahn, JG 54 under *Maj* Hannes Trautloft and JG 77.

On the first day of the campaign, JG 51 became the first *Geschwader* to claim its 1,000th victim, Mölders himself destroying five Russian aircraft. On 31 July 1941, JG 53 destroyed its 1,000th aircraft followed next day by JG 54, and on 15 August by JG 3.

Earlier in June 1941, 7./JG 26 under *Hptm* Joachim Müncheburg had been transferred to North Africa and in September, II./JG 27 joined the first *Gruppe*. In December 1941 III./JG 53 and III./JG 27 arrived in North Africa and I. and II./JG 53 were transferred to Sicily. On 24 January 1942 a completely new *Geschwader* was formed in Norway and Finland. Designated JG 5 *Eismeer* (Arctic sea), the unit was established from a nucleus provided by IV./JG 1 and I./JG 77. About the period JG 1 was brought up to full strength by the formation of new III. *Gruppe* and the redesignation of I./JG 3 to become II./JG 1.

At this time the majority of *Luftwaffe* fighter units were equipped with the Bf 109F, although the first Fw 190s began to reach JG 26 in August 1941. It was this unit, which at the time had some 30 Bf 109F–4s on hand combined with 90 similar aircraft from JG 2 and 60 from JG 1, that took part in Operation Thunderbolt. This was the code name for the break out of the battle cruisers *Scharnhorst* and *Gneisenau* from Brest to Kiel and Wilhelmshaven. During the operation, the *Luftwaffe* claimed 49 RAF aircraft destroyed.

On 10 November 1941 a special fighter-bomber *Staffel* had been formed within JG 2 under the command of *Hptm* Frank Liesendahl. The unit, which was designated 10.(Jabo)/JG 2 was equipped with the Bf 109F–4/B which could carry a 550 lb SC 250 bomb. The succes of the *Staffel* led to the establishment of a second unit designated 10.(Jabo)/JG 26 under *Hptm* Karl Plunser. Up until June 1942 when the units were re-equipped with the Fw 190A, 10./JG 2 claimed the destruction of 63,000 tons of shipping and 10./JG 26 carried out 32 successful raids on the south coast of England.

One of the most interesting developments of the Bf 109F was the H series of high altitude fighters. The Bf 109H V1 was fitted with an additional wing section which increased span to 40 ft 1½ ins and a braced tailplane. The aircraft, together with a small batch of Bf 109H–0s was powered by a 1,300 hp DB 601E engine. The production model, the Bf 109H–1 was issued to an operational reconnaissance unit in France, reported to have been 3.(F)/121. Several developments were proposed including the Bf 109H–2 and H–3 powered by 1,750 hp Jumo 213 engines and the H–5 with a DB 605L unit.

By the end of 1942 the Bf 109F had disappeared from service, giving way to the Fw 190 and the most numerous version of Messerschmitt's fighter, the Bf 109G—the 'Gustav'.

Specification (Bf 109F–3)

SPAN	9·92 m	(32 ft 6½ ins)
LENGTH	8·86 m	(29 ft 0¾ ins)
HEIGHT	3·40 m	(11 ft 2 ins)
EMPTY WEIGHT	1,960 kg	(4,321 lbs)
LOADED WEIGHT	2,750 kg	(6,063 lbs)
MAXIMUM SPEED	630 km/h	(391 mph)
NORMAL RANGE	710 km	(440 miles)

Top right: *The Bf 109G–0 pre-production aircraft was fitted with a DB 601E engine pending the arrival of the more powerful DB 605.*

Right: *The Bf 109G–1 could be distinguished from the F-series by the two tiny intakes aft of the spinner and the deletion of the triangular window just below the windscreen.*

Left: *The Bf 109F (w/nr 5603) which was fitted with a fixed nosewheel undercarriage for tests in connection with the Me 262 and Me 309 programmes.*

Bf 109G and K

The introduction of the Bf 109G into *Luftwaffe* service saw the start of a process in which Messerschmitt's excellent fighter was developed into a relatively slow and cumbersome anti-bomber weapon. Progressively more guns and armour were pressed upon the fighter with the result that it began to lose its parity with the Spitfire, and lose the confidence of its many devoted pilots.

The first Bf 109Gs were very similar in appearance to the previous F model, the main difference being the introduction of the more powerful 1,475 hp DB 605 engine. Externally, the two variants were very difficult to distinguish, perhaps the two major features being the introduction of two tiny intakes aft of the spinner and the deletion of the triangular windows below the windscreen side panels.

Twelve Bf 109G–0 pre-production aircraft were built, these retaining the 1,300 hp DB 601E engine. Provision was made for a pressurised cabin and the airframe was the subject of some structural strengthening. The first variant to be fitted with the DB 605A engine was the Bf 109G–1 which also employed GM–1 power boosting. This was a system whereby nitrous-oxide was injected into the supercharger to boost available power above the engine's rated altitude. The Bf 109G–1/Trop was fitted with tropical filters and carried two 13 mm MG 131 machine guns in place of the MG 17 weapons mounted above the engine cowling. The introduction of these weapons resulted in two large bulges forward of the cockpit, this gaining for the variant the nickname of the 'Beule' or 'Bump'.

The Bf 1209G– was fitted with a pressurised cabin and was intended mainly for the fighter-reconnaissance rôle. It reverted to the armament of the G–1, but could carry two rearward-firing MG 17 machine guns in a ventral pack. The G–2/R1 was a fighter-bomber with provision for an under-fuselage bomb and two underwing drop tanks. The aircraft was fitted with a third undercarriage leg to provide the necessary ground clearance, this being jettisoned by parachute after take-off. The Bf 109G–3 was similar to the G–1 but has FuG 16Z radio in place of FuG 7A and the G–4 was an unpressurised version of the G–3.

The Bf 109G–5 introduced the more powerful DB 605D engine with MW–50 injection. In this system, a mixture of water and methanol was injected into the engine

The Bf 109G–1/Trop was the first model to be equipped with the MG 131 machine gun. The bulges for these weapons forward of the cockpit are clearly visible on this photo.

cylinders to boost power to 1,800 hp for short periods. The variant carried the armament first introduced by the G–1/Trop, two MG 131 machine guns and a MG 151/20 cannon firing through the spinner. The Bf 109G–5/R2 was fitted with a new taller wooden tail and a lengthened tailwheel, it being hoped that the feature would help to cure the aircraft's take-off swing.

Without doubt the most important 'Gustav' was the Bf 109G–6 which first introduced the heavy weapons which were to become associated with the variant. In addition to the armament of the G–5, the new version had two extra MG 151/20 cannon mounted beneath the wing in gondolas. The armament proved very effective against Allied bombers, but severely reduced the aircraft's performance. The G–6/U4 carried two 30 mm MK 108 cannon in place of the underwing MG 151/20s, and the G–6/U4N carried special radar equipment with a rotating aerial aft of the cockpit for night fighting.

The Bf 109G–6/R1 was a fighter-bomber which could carry up to a 1,100 lb SC 500 bomb under the fuselage and the G–6/R2 was fitted with a 210 mm WGr 21 rocket tube under each wing. The Bf 109G–7 was a proposed variant which brought together all modifications introduced by the G–6 and the G–8 was a reconnaissance aircraft with either a Rb 12·5/7 or Rb 32/7 camera and reduced armament.

Late in May 1942 the first Bf 109G–1s were delivered to the newly formed 11 *Staffel* of JG 2 to operate high altitude sorties. During the autumn of the year the Bf 109G began to enter service with many other units including JG 1, JG 2, JG 26 and JG 27. In September a Bf 109G–2 (w/nr 14256) was delivered to *Hptm* Hans-Joachim Marseille, who at that time was commanding 3./JG 27. Marseille had already proved himself to be one of the most outstanding pilots of the war, it not being unusual for such pilots as Homuth (commander of I./JG 27) and Franzisket to break off combat to watch his displays. After claiming 151 victories, Marseille was lost on 30 September 1942. His aircraft sprung a glycol leak and the fumes forced the young pilot to bale out. In so doing, his body struck the tailfin, and he was knocked unconscious being unable to pull his parachute ripcord.

By the beginning of 1943, the Bf 109G had virtually replaced the F series in *Luftwaffe* service. JG 77 under *Maj* Müncheburg was transferred to North Africa, eventually replacing JG 27 under *Oberstlt* Neumann which was withdrawn via Greece and Sicily to Germany. JG 53 under *Maj* von Maltzahn was based in Tunisia and II./JG 51 under *Hptm* Grasser was transferred to the area from Russia. II./JG 2 also saw service in Tunisia for a short period.

The remainder of JG 2 and JG 26 which were on the Channel Coast were still equipped in the main with the Fw 190, but JG 1 under *Oberstlt* Mix in Germany used both Fw 190s and Bf 109Gs. JG 54 under *Maj* Trautloft, which was based in northern Russia, operated a mixed bag of Bf 109Gs and Fw 190s. III./JG 54 under *Maj* Seiler with Bf 109Gs was transferred to the west, and was joined soon afterwards by a fourth *Gruppe*. I. and II./JG 51 were in central Russia with the Fw 190, whilst in the Crimea the *Luftwaffe* had the Bf 109Gs of JG 3 under *Oberstlt* Wilcke and JG 52 under *Oberst* Hrabak. One other *Jagdgeschwader* was operational, JG 5 under *Oberstlt* Handrik based in Norway and Finland.

The spring of 1943 saw an increase in the daylight raids carried out by the US Eighth Air Force, and to combat these the fighter units in Germany were considerably strengthened. III. and IV./JG 54 had been added to the strength of JG 1, and on 1 April 1943 a new *Geschwader* was created from a nucleus provided by the latter unit known as JG 11.

Several new weapon combinations were developed during this time in an effort to quickly destroy the heavily armoured B–17 and B–24 bombers used by the USAAF. One very novel scheme was the attaching of a 550 lb bomb beneath the fuselage of a Bf 109G, the pilot dropping this missile on top of the bomber fleet. This scheme was projected by *Lt* Heinz Knocke of 5./JG 1 who first destroyed a B–17 using the weapon on 22 March 1943. Aircraft from JG 1 and JG 26 and later JG 3 were equipped with 210 mm rocket tubes and least one Bf 109 was tested with four 50 mm RZ 65 rocket tubes mounted in each wing.

On 14 May 1943, III./JG 54 lost eight Bf 109Gs, plus a further seven on the next day. This occurred during an interception of 300 four-engined bombers over the German Bight, the *Gruppe* losing one of its leading pilots, *Lt* Rupp with 50 victories, near Heligoland. During this period however it was the *Luftwaffe* that did most of the scoring. On 28 July 1943, *Fw* Fest of 5./JG 11 destroyed three B–17s with a direct hit from a 550 lb bomb. Of the 22 USAAF bombers lost on this date, no less than half were claimed by *Hptm* Specht of II./JG 11. Specht had lost an eye very early in the war, but returned to combat to become one of the outstanding pilots operating in the defence of Germany.

During the first day of August 1943, USAAF bombers attacked the Rumanian oilfields at Ploesti. They were intercepted by the newly formed I./JG 4 under *Hptm* Hahn based at Mitzil and harried by IV./JG 27 under *Oblt* Burk based at Kalamaki. The USAAF lost a total of 54 aircraft during the raid.

One of the most important actions by *Luftwaffe* fighters on the eastern front began on 5 July 1943 when Germany launched Operation Zitadelle. This was the greatest tank battle that the world has ever seen, and the last major German offensive in the east. In addition to a large number of ground attack units, II. and

Right: *Thirty Bf 109G–2s were delivered to the Finnish Air Force and used with great success against Russian forces.*

III./JG 3, I., III. and IV./JG 51, I. and III./JG 52 and a part of JG 54 were operational. The fighter arm claimed 432 victories during the first day of the campaign, including 77 by II./JG 3 flying Bf 109Gs. Highest scorers during the day were *Hptm* Wiese with 12, *Oblt* Krupinski with 11, both pilots coming from III./JG 52 and *Oblt* Kirschner with 9 from II./JG 3.

Five *Jagdgruppen* were transferred to the defence of the Reich late in July 1943 including I., II. and III./JG 3 from Russia, II./JG 27 from Italy and II./JG 51 from Sardinia. These units were to prove their worth when 60 out of a force of 376 bombers attacking the Schweinfurt ball-bearing factories were destroyed and a further 100 damaged. In a second attack on 14 October the USAAF lost a further 60 bombers from a force of 291.

On 27 July 1943 *Maj* Hajo Herrmann, a young bomber pilot, put forward a new method of attacking the RAF bomber raids which were ceaselessly pounding German cities. His *Wilde Sau* (Wild Boar) tactics as they were subsequently designated, involved the use of free-lance single-engined fighters to strike at the bombers over the target making use of the glow from searchlights, flak and flares. Goring was enthusiastic to the proposal and authorised the establishment of an experimental unit known as *Kommando Herrmann*.

The successes of the unit were such, that the establishment of a *Wilde Sau Geschwader* was ordered under the designation

A captured Bf 109G–2/Trop is examined. This machine probably belonged to 2./JG 27, the most famous Geschwader *to operate the Bf 109 in North Africa.*

The Bf 109G–2/R1 with its rather curious third undercarriage leg. The variant could carry two 300 litre drop tanks and a 550 lb bomb.

JG 300 with Herrmann at its head. Continued successes during the light summer nights of 1943 led to the establishment of two further units; JG 301 under *Maj* Weinrich and JG 302 under *Oberstlt* Janssen. Only one *Gruppe* in each *Geschwader* possessed its own aircraft, the other *Gruppen* sharing their machines with a day fighter unit based on the same airfield. The constant use of aircraft had a drastic effect on serviceability rates, and following severe losses during the winter 1943–44, many of the units transferred to conventional day fighter operations.

About this time, the American escort fighters began to appear in greater numbers, and some were even able to stay with the bombers throughout their sorties. Many *Luftwaffe* aces lost their lives during this period. *Oberstlt* Phillipp of JG 1 was lost on 8 October 1943 to P–47s of the USAAF, *Oberstlt* Mayer of JG 2 was killed on 2 March 1944 also in combat with P–47s, *Oberst* Wilcke of JG 3 was shot down by P–51s on 23 March and *Oberst* Oesau of JG 1 was killed on 11 May following combat with P–38s.

Just after this, on 20 May 1944, *Maj* Dahl of III./JG 3 was ordered to report to *General* Galland who was then Inspector of Fighters. Galland asked him if he would form and command a new *Geschwader* to be based in the Nuremberg-Ansbach area comprising five existing fighter *Gruppen*. Dahl was very enthusiastic about the idea, and two days later the *Jagdgeschwader zbV* had been formed. It comprised III./JG 3 now under *Hptm* Langer, I./JG 5 under *Maj* Carganico, II./JG 27 under *Hptm* Franzisket, II./JG 53 under *Hptm* Meimberg and III./JG 54 under *Hptm*

Schroer, all equipped with Bf 109Gs. The unit proved quite successful, although *Maj* Carganico flying a Bf 109G–5/U2 (w/nr 110087) was killed on 27 May. Following the invasion of Europe, the *Gruppen* returned to their parent *Geschwader* and the *Stab*/JGzbV became the *Stab* of JG 300.

The fastest of all 'Gustavs' was the Bf 109G–10 which was powered by a DB 605D engine with MW–50 injection. The variant carried an armament of two 13 mm MG 131 machine guns and a 30 mm MK 108 cannon firing through the spinner. It could attain a speed of 428 mph at 25,000 ft and reach an altitude of 20,000 ft in 6 minutes. The Bf 109G–10/U4 carried two MK 108 cannon in a belly tray which could be replaced by a non-jettisonable fuel tank known as the *Irmer Behalter*. The Bf 109G–10/R2 and R6 were fitted with the taller wooden tail surfaces introduced by the G–5/R2 and many G–10 variants were fitted with the clear-blown *Vollsichthaube* canopy. This was often used in conjunction with a piece of 90 mm armoured glass (known as the *Kopfschutz Galland*) to afford the pilot a much improved view. The feature was fitted to many late production Bf 109Gs and Ks.

The Bf 109G–12 was a tandem two-seat trainer based on the G–1. A second seat for the instructor was mounted behind the normal pilot's position, the whole enclosed by a new glazed canopy. Most Bf 109G–12s were in fact converted from standard Gustavs by *Luftwaffe* maintenance units and served with the training *Jagdgeschwader* such as JG 101, 102, 104, 106, 107 and 108.

Final variant of the G–series to see operational service was the Bf 109G–14 fighter-bomber which was fitted with the *Vollsichthaube* canopy as standard and carried an armament of one MG 151/20 cannon and two MG 131 machine guns. Provision was made for two extra MG 151/20s, two WGr 21 *Dodel* rocket tubes or a 550 lb bomb. The G–14/R2 was fitted with the taller wooden tail assembly and the G–16 was heavily armoured ground-attack variant which probably failed to see service.

The Bf 109H has been previously described, the designation Bf 109I was not used and the Bf 109J was the German description for a variant to be built in Spain. The next, and last version to see operational service was the Bf 109K. This was basically similar to the Bf 109G–14 but featured the deletion of the 'bumps' which had characterised all Bf 109s from the G–1/Trop.

The Bf 109K–0s were in fact converted

The Bf 109G–10/R2 was the first variant of the fighter to be equipped with the 'Vollsichthaube' *clear-blown cockpit hood. This aircraft was operated by JG 54* Grunherz.

Top left: *The Bf 109G–6/R2 carried two 210 mm WGr 21 rocket tubes underwing.*

Left: *The Bf 109G–6/R6 carried two 20mm MG 151 cannon beneath its wings.*

Right: *Apart from the Bf 109G–2 and G–6 variants, a number of Bf 109G–14s were also delivered to the Finnish Air Force.*

Top left: Luftwaffe *maintenance units converted many G-series aircraft into Bf 109G–12 two-seat trainers.*

Left: *At least one Bf 109G was fitted with four RZ 65 rocket barrels in each wing.*

from standard Gustavs, being powered by the DB 605D engine with GM–1 power boost. The Bf 109K–2 and K–4 were similar but introduced the 1,550 hp DB 605ASCM/DCM engine, which with MW–50 injection, could develop 2,000 hp for short periods. Both variants carried an armament of a 30 mm MK 103 or MK 108 and two 15 mm MG 151 cannon, the sole difference between the two being that the K–4 had a pressurised cabin. The Bf 109K–6 was similar to the K–2 but carried an armament of two MG 131 machine guns and three of the long-range MK 103 cannon. The final variant was the Bf 109K–14 which carried a similar armament to the K–2, but was powered by the DB 605L engine which gave it a maximum speed of 450 mph.

As can be imagined, a large variety of experimental variants of the Bf 109 were projected or built. Included amongst these were the Bf 109L which was to be powered by a 1,750 hp Jumo 213 engine, with which it was hoped to reach a speed of 474 mph. The Bf 109S was a proposed variant with blown flaps to improve low-speed characteristics. This feature was tested by the Bf 109 V24 in the Chalais-Meudon wind tunnel. Another interesting variant was the Bf 109Z which employed two Bf 109F airframes coupled together by a new centre section.

Other Bf 109s were fitted with such features as a nosewheel undercarriage, a butterfly tail, a BMW 801 radial engine for comparison with the Fw 190, and even a Rolls-Royce Merlin engine for tests with a captured Spitfire. Perhaps the most interesting project however was the Bf 109TL.

Developed in January 1943, the Bf 109TL was basically the fuselage of a Bf 109 married to the wings of the Me 155 project and utilising the nosewheel from the Me 309. The design was to be powered by two Junkers Jumo 004B turbojets and was intended to produce a jet fighter that would cause less disruption to production than the introduction of the entirely new Me 262. Although quite feasible, the Bf 109TL was abandoned after two months when it was realised that the Me 262A–0 would soon be leaving the assembly lines. The project was estimated to have a speed of 522 mph and a ceiling of 37,500 ft.

When the Allies invaded Normandy on 6 June 1944, only two fighter units were operational on the western front, JG 2 under *Oberstlt* Bühlingen and JG 26 under *Oberst* Priller. During the next few days no less than 23 *Jagdgruppen* were brought into the area including 13 with Bf 109Gs. These were III./JG 1, II./JG 2, II. and III./JG 3, I. and II./JG 5, II./JG 11, I., II. and IV./JG 27, II./JG 53 and I./JG 301.

In the east too, the *Luftwaffe* was facing desperate odds. On 23 August 1944, a coup d'etat in Rumania coupled with a surprise Russian offensive led to the withdrawal of German forces. Bf 109Gs from I./JG 53 under *Maj* Harder and II./JG 301 under *Hptm* Jacobs helped to cover the retreat. In Hungary, the *Stab* of JG 76 under *Maj* Düllberg was transferred to take over the control of the *Gruppen*

An extremely rare photo of the Bf 109G which was experimentally fitted with a butterfly tail. The aircraft possessed better flight and performance characteristics than the standard aircraft but was not accepted.

previously assigned to *Oberstlt* Steinhoff's JG 77. These were II./JG 51 under *Maj* Rammelt, II./JG 52 under *Maj* Barkhorn and I./JG 53 under *Maj* Hardu.

Before leaving this theatre of operations, mention must be made of two *Luftwaffe* fighter pilots: *Hptm* Erich Hartmann who eventually commanded I./JG 52 and *Maj* Gerhard Barkhorn who at one stage of the war led JG 6. Both pilots flew Bf 109s throughout the war, Hartmann claiming the fantastic number of 352 victories, and Barkhorn, 301. Even when Barkhorn joined the Fw 190D–9 equipped JG 6, he and his wingman retained their Bf 109G–14s. Another great German pilot was *Oberst* Hermann Graf who, at the end of the war, commanded the *Luftwaffe*'s top scoring fighter unit, JG 52 with almost 12,000 kills. Graf, who flew the Bf 109 throughout his career, claimed 212 victories.

During August 1944, a number of new fighter units were established concurrently with the disbandment of virtually all bomber groups. I./JG 4 was joined by three more *Gruppen*, II./JG 4 being formed from the *Ramm Staffel*, and IV./JG 4 from II./JG 5. A completely new *Geschwader*, JG 6, was established under *Oberst* Kogler. I. and II./JG 6 were formed from I. and II./ZG 26 and III./JG 6 was renamed from I./JG 5. Finally, II./ZG 1 became III./JG 76, but this unit lasted for a month before it became IV./JG 53.

The *Luftwaffe*'s last major operation came on 1 January 1945 when 750 fighters straffed Allied airfields in Holland, Belgium and Luxembourg. Code named Operation Bodenplatte, the German fighters, which included a large proportion of Bf 109s, claimed the destruction of 137 British and 115 American aircraft. *Luftwaffe* losses were in the region of 200, some of them, including *Oberstlt* Specht of JG 11, coming from experienced leaders who were difficult to replace.

From then on, *Luftwaffe* operations steadily declined. Allied bombers began a concentrated attack on German fuel supplies and many fighters were unable to get off the ground. The last major

A number of Messerschmitt Bf 109G–14s were built under licence by the Czech Avia Company. This machine, OK–BYH, had a red nose and wing tips and was operated by the Czech Air Police.

operation by Bf 109s was carried out at the instigation of *Oberst* Hajo Herrmann. A special fighter unit of four *Gruppen* was established known as *Rammkommando Elbe*. Most of the pilots were students with perhaps one or two solo flights to their credit and all were volunteers whose enthusiasm was bolstered with martial music and patriotic speeches. Only one operation was undertaken by the unit on 7 April 1945 when 120 Bf 109Gs took-off to intercept a USAAF bomber formation. Only 15 returned.

Apart from use by the *Luftwaffe* aircraft were also delivered to Bulgaria (19 Bf 109Es and 145 Bf 109Gs), Croatia (73 Bf 109Es and a small number of Bf 109Gs), Slovakia (a few Bf 109Es and Gs), Hungary (49 Bf 109Es, 59 Bf 109Fs and Gs and about 700 nationally-built Bf 109Gs), Rumania (29 Bf 109Es and 70 Bf 109Gs) and Switzerland (10 Bf 109Ds, 80 Bf 109Es and 12 Bf 109G–6s). Aside from the *Luftwaffe* it was perhaps with the Finnish Air Force that the Bf 109 was most notable. Thirty Bf 109G–2s and 132 Bf 109G–6s were delivered to that air arm, entering service with HLeLv 34 later in 1943. The top scoring Finnish ace was Juutulainen with 94 victories.

Specification (Bf 109K–4)

SPAN	9·92 m	(32 ft 6½ ins)
LENGTH	9·02 m	(29 ft 7 ins)
HEIGHT	3·40 m	(11 ft 2 ins)
LOADED WEIGHT	3,370 kg	(7,438 lbs)
MAXIMUM SPEED	728 km/h	(452 mph)
NORMAL RANGE	590 km	(366 miles)
SERVICE CEILING	12,500 ft	(41,012 ft)

Bf 109 fighter units September 1939

I./JG	1	Jesau
I./JG	2	Döberitz
10./JG	2	Döberitz
Stab/LG	2	Jüterbog
I./LG	2	Garz
Stab/JG	3	Bernburg
I./JG	3	Zerbst
I./JG	20	Döberitz
I./JG	21	Jesau
I./JG	26	Köln
II./JG	26	Düsseldorf
11./JG	26	Köln-Ostheim
I./JG	51	Aibling
I./JG	52	Böblingen
I./JG	53	Wiesbaden
II./JG	53	Mannheim
I./JG	70	Herzogenaurach
1./JG	71	Böblingen
2./JG	71	Fürstenfeldbruck
I./JG	76	Wien-Aspern
I./JG	77	Breslau
II./JG	77	Pilsen
II./ZG	1	Fürstenwalde
I./ZG	2	Bernburg
I./ZG	26	Dortmund
II./ZG	26	Werl
III./ZG	26	Lippstadt
I./ZG	52	Illesheim
II./ZG	76	Gablingen

Bf 109 fighter units December 1944

III./JG	1	Anklam
II./JG	2	Nidda
Stab/JG	3	Störmede
I./JG	3	Werl
III./JG	3	Schachten
I./JG	4	Darmstadt
III./JG	4	Ebelsbach
IV./JG	4	Rhein-Main
Stab/JG	5	East Norway
III./JG	5	East Norway
IV./JG	5	Norway
III./JG	6	Schwerin
II./JG	11	Hannover
III./JG	26	Plantlunne
Stab/JG	27	Hopsten
I./JG	27	Rheine
II./JG	27	Hopsten
III./JG	27	Hesepe
IV./JG	27	Achmer
Stab/JG	51	Poland
I./JG	51	Poland
II./JG	51	Poland
III./JG	51	Poland
IV./JG	51	Poland
Stab/JG	52	Poland
I./JG	52	Poland
II./JG	52	Balkans
III./JG	52	Poland
Stab/JG	53	Donaueschingen
I./JG	53	Balkans
II./JG	53	Huchenfeld
III./JG	53	Neuthausen Ob Eck
IV./JG	53	Donaueschingen
Stab/JG	76	Balkans
Stab/JG	77	Schonwald
I./JG	77	Schonwald
II./JG	77	Eggersdorf
III./JG	77	Neuruppin
I./JG	300	Borkheide
III./JG	300	Jüterbog
IV./JG	300	Reinsdorf

(*III. and IV./JG* 5 *were also equipped with the Fw* 190. *All other* Jagdgruppen, *with the exception of JG* 400 *with the Me* 163, *were equipped with the Fw* 190.)

Bf 110C, D, E, and F

'THE Bf 110 and you who fly it will be like Hannibal's cavalry protecting the elephants; the bombers are my elephants.' Göring's words graphically describe the rôle for which the Bf 110 was conceived, that of long-range escort fighters imaginatively known as *Zerstören* (destroyers). In the event, the reputation of the Messerschmitt fighter was lost in the summer of 1940 when it was decimated by RAF Hurricanes and Spitfires; but it was only in its designed rôle that it had failed. During the mid-war years the Bf 110 was to provide the *Luftwaffe* with an indispensable interceptor to combat the ever more damaging raids of the USAAF by day and the RAF by night.

The aircraft was conceived late in 1934 as a result of an RLM specification for a long-range strategic fighter. It was designed around two of the new twelve cylinder DB 600A engines which were intended to develop 1,000 hp. Work on the first prototype, the Bf 110 V1, began at the Bayerische Flugzeugwerke's Augsburg plant, the machine making its maiden flight on 12 May 1936. Two further prototypes followed, the Bf 110 V2 which flew on 24 December 1936, and the V3 which was delivered on 14 January 1937.

Although lacking in manoeuvreability, the Bf 110 possessed a maximum speed comparable with that of the RAF's new Hurricane fighter; and this prompted the RLM to place a contract for four Bf 110A–0s. These were delivered between August 1937 and March 1938 and differed in being powered by two 670 hp Jumo 210B engines and carrying four 7·9 mm MG 17 machine guns in the nose and a MG 15 of similar calibre in the rear of the cockpit.

The underpowered Bf 110A was quickly abandoned in favour of the B-series, although both Bf 110B–0s were powered by Jumo 210 units. The Bf 110B–1 reverted to the more powerful DB 600A engines which were then being delivered in numbers. Two of the MG 17 machine guns were replaced by two 20 mm MG FF cannon,

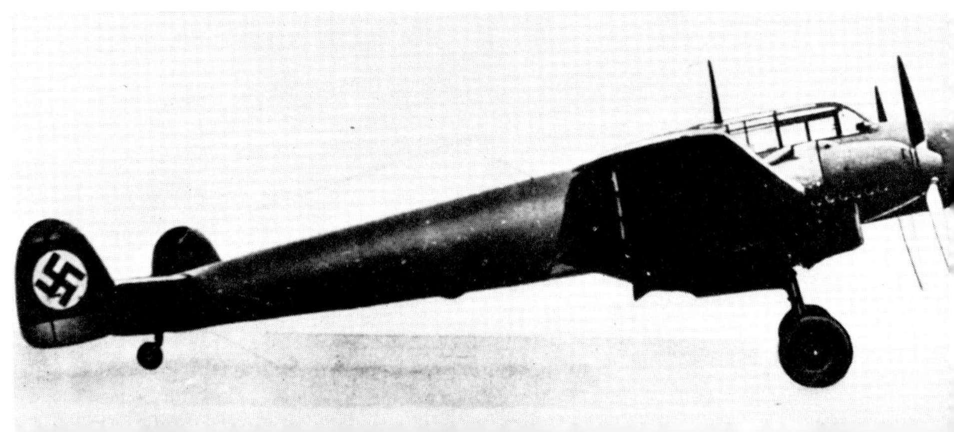

Top: *The Bf 110 V1 first flew on 12 May 1936 from Messerschmitt's Augsburg plant.*

Right: *Only a few Bf 110A–0s were delivered before production switched to the Bf 110B.*

A pair of operational Bf 110C–1s photographed just before the Second World War, possibly belonging to I.(Z)/LG 1.

but relatively few B–1 and B–2 fighters and B–3 trainers were completed before production switched to the Bf 110C.

The Bf 110C–0, which began to reach the *Luftwaffe* in February 1939, was powered by two 1,100 hp DB 601A engines with direct fuel injection. The Bf 110C–1 production model was similar, the C–2 introduced FuG 10 and FuG 111aU radio equipment, and the C–3 and C–4 were respectively similar apart from having improved MG FF cannon.

Early in 1939, Göring authorised the formation of a number of *Zerstörergruppen*, with the intention of equipping them with the Bf 110. Most of these *Gruppen* were formed from the *Luftwaffe*'s most famous *Geschwader*, JG 132 and JG 134. By the time the war began in September 1939 it had proved possible to equip only three of the nine *Gruppen* with the Bf 110. These were I.(Z)/LG 1 at Barth under *Maj* Walter Grabmann, I./ZG 1 at Jüterbog-Damm under *Maj* Joachim Friedrich Huth and I./ZG 76 at Olmütz under *Hptm* Reinecke. The remaining *Gruppen*, II./ZG 1, I./ZG 2, I., II. and

III./ZG 26, I./ZG 52 and II./ZG 76 retained their interim equipment of Bf 109Cs and Ds.

All three Bf 110 equipped *Gruppen* took part in the Polish campaign, *Maj* Grabmann being injured in combat with PZL P.11c fighters on 1 September 1939. Aircraft from I./ZG 76 saw combat with RAF Wellington bombers during December, but it was not for four months that the *Luftwaffe* was again involved in a major action. The Bf 110 was ideally suited to the long distances involved in Operation Weserübung, the action against Denmark and Norway which began on 9 April 1940. Two units were assigned, I./ZG 1 at Barth and I./ZG 76 at Westerland, the units transferring to Aalborg and Stavanger by the evening.

The operations in Norway continued until May 1940 when I./ZG 1 was withdrawn to Germany. On 10 May the *Luftwaffe* attacked France and the Low Countries, the force having the same number of destroyer *Gruppen* with which it began the conflict, all now equipped with the Bf 110. The *Zerstören* were engaged mainly in ground support operations, but on 1 June 1940, ZG 26 tangled with RAF Hurricanes and Spitfires for the first time over Dunkirk.

*A Bf 110C–2 (M8 + CP) of 6.*Staffel*/ZG 76 in flight. Note the* Haifisch *(shark's teeth) insignia on the nose.*

Meanwhile several other variants of the Bf 110 had been produced. The C–1/U1 was a glider tug, the C–2/U1 was fitted with remotely-controlled gun barbettes and the C–4/B was a fighter-bomber. The Bf 110C–5 was a reconnaissance aircraft which was first delivered to 1./Aufkl.Gr. Ob.d.L. and 3. and 4.(F)/11. The Bf 110C–6 was fitted with a 30 mm MK 101 cannon and the C–7 could carry two 1,100 lb bombs.

During the spring of 1940 the first Bf 110D–0s were delivered to V.(Z)/LG 1. This variant had the MG FF cannon removed and carried a huge 230 gallon fuel tank beneath the fuselage. The D–1/R1 carried two jettisonable tanks beneath the wings of up to 198 gallon capacity and the D–1/R2 could mount all three tanks. The D–1/U1 and proposed D–4 production model were experimental night fighters, the D–2 was a fighter-bomber and the D–3 could carry both bombs and drop tanks.

Late in June 1940, I./ZG 1 under *Maj* Wolfgang Falck was withdrawn to Dusseldorf and redesignated initially as I.(*Nacht*)/JG 1 and eventually as I./NJG 1. The unit was formed in an attempt to combat the growing RAF night raids with which the Flak units were unable to cope. *Maj* Falck was promoted to command NJG 1 and *Hptm* Günther Radusch took over I./NJG 1, a 4th and 5th *Staffel* being added from *Zerst.Sta.*/KG 30 and a special Do 17 fighter squadron.

On 6 July 1940, I./ZG 52 was redesignated as II./ZG 2 and II./ZG 1 became III./ZG 76. For the Battle of

Top: *A Bf 110C–5 photographic reconnaissance aircraft of 2.(H)/14 in North Africa. The* Luftwaffe *ground crew in the foreground are working on the aircraft's camera*

Left: *A huge auxiliary fuel tank experimentally fitted to the Bf 110C.*

Top: *1./ZG 76 operated the Bf 110D–0 with its huge auxiliary fuel tank during the campaign against Norway and Denmark in April and May 1940. This machine was coded M8+FH and carried the Werke Number 3127.*

Right: *The Bf 110D–3 carrying two 300 litre drop tanks beneath the wings. Just outboard of the attachment points can be seen the four ETC 50 bomb racks.*

Left: *The Bf 110C–7 fighter-bomber in flight over a barren Arctic coastline.*

Britain the following Bf 110 *Gruppen* were operational: I./ZG 2 under *Maj* Ott, II./ZG 2 under *Maj* Carl, I./ZG 26 under *Hptm* Mackrocki, II./ZG 26 under *Hptm* von Rettburg, III./ZG 26 under *Hptm* Schalk, I./ZG 76 under *Hptm* Restemeyer, II./ZG 76 under *Hptm* Groth, III./ZG 76 under *Hptm* Dickore, V.(Z)/LG 1 under *Hptm* Leinsberger and *Erprobungsgruppe* 210 under *Hptm* Rubensdorffer. All, apart from I./ZG 76 at Stavanger, were based in northern France.

The first major action in which the Bf 110 was involved came on 11 August when 61 aircraft from ZG 2 escorted bombers from KG 27 and KG 54. They were intercepted by eight RAF squadrons and 11 Bf 110s were shot down, including that of *Maj* Ott, commander of I./ZG 2. Altogether the *Luftwaffe* lost 17 Bf 110s. Next day 11 Bf 110s were lost, and on 13 August, 20 aircraft from V.(Z)/LG 1, ZG 2, ZG 26 and ZG 76 were shot down. The *Kommodore* of ZG 26 was *Oberstlt* Joachim Huth, a veteran pilot of the First World War who had a wooden leg.

Without doubt the most disastrous losses occurred on 15 August when 27 Bf 110s were destroyed, 19 of them from ZG 76, and 7 from *Erprobungsgruppe* 210. It was perhaps the last named unit, practising fighter-bomber tactics, that suffered the most during the daylight battles. Between 4 August and 5 October 1940, they lost 30 Bf 110s and three commanders, *Hptm* Rubensdorffer on 15 August, *Hptm* von Boltenstern on 4 September and *Hptm* Lutz on 27 September. On the latter day the *Luftwaffe* lost 20 Bf 110s.

The loss of over 300 Bf 110s during the Battle of Britain confirmed that the unwieldy machine was unable to meet the very manoeuvreable fighters of the RAF on anything like equal terms.

During the autumn of 1940 the build up of the *Luftwaffe*'s new night fighter arm continued. III./NJG 1 was formed from IV.(N)/JG 2 in July 1940 and on 11 September II./NJG 1 was redesignated as I./NJG 2. A second II./NJG 1 was formed from I./ZG 76 under *Hptm* Graf von Stillfried on 2 October and in the same month, V.(Z)/LG 1 became I./NJG 3, *Hptm* Radusch taking over command.

On 30 January 1941 III./ZG 26 under *Maj* Kaschka became the first unit to arrive in North Africa when it was transferred to Castel Benito. The *Luftwaffe*'s next major action was the assault on the Balkans. Amongst the attacking forces were Stab, I. and II./ZG 26 under *Oberstlt* Schalk equipped with the Bf 110. During 22 May 1941 Royal Navy units were reported to be in positions north and west of Crete. ZG 26's Bf 110s based at Argos in company with Bf 109s escorted the bombers and dive bombers of VIII Fliegerkorps as they attacked the British ships. The cruisers HMS *Gloucester* and HMS *Fiji* were sunk, and in the next few days several destroyers went to the bottom of the sea.

Right: *The specialised Bf 110E-1 fighter-bomber in flight. The bomb racks mounted under the wings and fuselage centre section are clearly visible.*

The Bf 110E–1/U2 was powered by two DB 601N engines. This machine was operated by III./ZG 1 during 1943.

About this time the Bf 110E began to enter service. The E–0 and E–1 were similar to the C–4/B, but could carry four 110 lb bombs beneath the wings in addition to the 550 lb weapon beneath the fuselage. The Bf 110E–1/Trop was fitted with desert survival equipment, the E–1/U1 was an experimental night fighter and the E–1/U2 was powered by two 1,200 hp DB 601N engines. The E–1/R2 could carry two 2,200 lb bombs and the E–2, originally known as the E–1/R3 was similar but powered by DB 601N engines. The final E-series aircraft was the Bf 110E–3 fighter-reconnaissance machine with reduced armament and two 198 Imperial gallon drop tanks.

At 03.15 hours on 22 July 1941, the *Luftwaffe*, in co-operation with three Army Groups, launched Operation Barbarossa—the invasion of the Soviet Union. It is some indication of the esteem in which the Bf 110 was held by the German Air Force that only one destroyer *Geschwader*, ZG 26 *Horst Wessel* remained, In addition a new unit, SKG 210, was operational, its two *Gruppen* having been formed from Erpr.Gr 210 and III./ZG 76 in April 1941.

Five *Nachtjagdgruppen* (night fighter groups) were active at this time; I., II. and III./NJG 1, I./NJG 3 and I./NJG 2, all but the latter equipped with the Bf 110. During October and November 1941 three further *Gruppen* were formed; II./NJG 2 under *Oblt* Lent, II./NJG 3 under *Maj* Radusch (*Hptm* Knoetzsch taking over I./NJG 3) and III./NJG 3 under *Hptm* Nacke.

Initially the night fighters attacked visually, but this hit and miss method was soon replaced by a system of ground-controlled interception using the giant Würzburg radar sets for guidance. This chain of radar stations, which eventually stretched from Denmark to the Swiss border, was given the code name *Himmelbett*, but was popularly known as the 'Kammhuber Line' after the chief of the *Luftwaffe*'s night fighter arm.

The first variant of the Bf 110 to be built specially for night fighting was the F–4. The previous Bf 110F–0, F–1 and F–3 were similar to the E–0, E–1 and E–3 apart from the introduction of 1,300 hp DB 601F engines, and the F–2 was a bomber-interceptor with provision for four 210 mm WGr 21 rocket tubes underwing. The new Bf 110F–4 carried a nose armament of four MG 17 machine guns, two MG FF and two 30 mm MK 108 cannon, plus a third crew member. The F–4/U1 was fitted with two 20 mm cannon firing obliquely upwards in the so-called *Schräge Musik* installation.

Specification (Bf 110C–4)

SPAN	16·20 m	(53 ft 4¾ ins)
LENGTH	12·10 m	(39 ft 8½ ins)
HEIGHT	3·50 m	(11 ft 6 ins)
EMPTY WEIGHT	5,200 kg	(11,466 lbs)
LOADED WEIGHT	6,750 kg	(14,884 lbs)
MAXIMUM SPEED	560 km/h	(349 mph)
NORMAL RANGE	775 km	(481 miles)

Right: *This Bf 110C–4 crashed at Prague-Rusin airfield on 12 September 1944.*

Bf 110G

By the end of 1941 it was clear that the long awaited Me 210, intended to replace the Bf 110, was not a success. To fill the gap thus created, a 'cleaned-up' version of the Bf 110 was introduced powered by two 1,475 hp DB 605B engines and designated Bf 110G. The Bf 110G–0 and G–1 were basically similar to the E–0 and E–1, but the G–2 was a fighter-bomber with a reinforced undercarriage and the two 20 mm MG FF cannon replaced by the much improved MG 151/20 weapons of similar calibre. Two 1,100 lb and four 110 lb bombs could be carried externally.

The Bf 110G–2/R1 differed in carrying a 37 mm Flak 18 gun beneath the fuselage, the G–2/R2 was a heavy fighter with the rear gun replaced by a fuel tank for use in connection with the GM–1 power boosting system. The G–2/R3 carried a forward firing armament of four MG 151/20 and two MK 108 cannon, the G–2/R4 differed solely from the G–2/R1 in having MK 108 cannon in place of the MG 17 machine guns and the similar G–2/R5 carried GM–1 power boosting.

The Bf 110G–3 was a high speed reconnaissance aircraft with armament reduced to four MG 17 machine guns, the G–3/R3 carrying two MK 108 cannon. The most important production model was the Bf 110G–4 night fighter which introduced airborne radar for the first time. In July 1941 the first such set, the *Lichtenstein* BC underwent trials, being introduced into service in February 1942. By the autumn the majority of German night fighters were equipped with either *Lichtenstein* BC or the simplified C–1 version.

The Bf 110G–4a and G–4b both carried FuG 212 *Lichtenstein* C–1 radar, but the latter also mounted FuG 220b *Lichtenstein* SN–2. The Bf 110G–4c carried FuG 220b only and the G–4d, of which 45 were built in March 1945, employed an improved aerial array and at least one aircraft was fitted with FuG 227 *Flensburg* 'passive' radar. No less than eight U-series modifications and R-series conversion packs were produced for the Bf 110G–4, mainly involving alterations in armament, fuel tankage and the provision of GM–1 power boosting.

The Bf 110F–4 and G–4 were produced for some time in parallel and by the end of 1942 the *Luftwaffe*'s night fighter arm comprised 389 aircraft of which the majority, at least 300, were Bf 110s. During the year the RAF lost 1,291 bombers in night attacks over occupied Europe, an estimated two-thirds of these to night fighters. For a 'failure' the Bf 110 was proving remarkably successful.

By early 1943, only one Bf 110 destroyer *Gruppe* remained, III./ZG 26 under *Maj* Cristl based in Sicily. In contrast there were 16 Bf 110 night fighter *Gruppen* operational, four in NJG 1, NJG 3, NJG 4 and NJG 5. I. and II./NJG 2 were to retain their Ju 88s until the end of the war.

The *Himmelbett* system proved very effective until 24 July 1943 when the RAF attacked Hamburg. The raid was reported as it passed north of the Elbe estuary and NJG 3's Bf 110s were scrambled to intercept. Suddenly the Würzburg radar screens were jammed by a mass of traces—the RAF was using 'Window' for the first time. This was a series of metal foil strips dropped from each aircraft which effectively rendered the German radar useless.

The *Luftwaffe* produced two temporary solutions to the problem of 'Window'. The first was code-named *Wilde Sau* (Wild Boar) and involved the night fighters intercepting the bombers over the target itself, using the light from flares and searchlights to attack visually. *Zahme Sau* (Tame Boar) tactics, pioneeerd by *Oberst* von Lossberg, involved the night fighters infiltrating into a bomber stream before or after it reached the target.

In addition to the powerful forms of armament previously described, many Bf 110s were fitted with the *Schräge Musik* (Jazz Music) installation. This comprised two 20 mm cannon arranged to fire forwards and upwards at an angle between 70 and 80 degrees and aimed by means of a special reflector sight.

Schräge Musik was invented by an armourer, Paul Mahle. While at the armament test centre at Tarnewitz, Mahle saw a Do 217 fitted with obliquely firing guns as defence against fighter attack. When Mahle joined II./NJG 5 he mounted two MG FF cannon behind the rear bulkhead of a Bf 110 to fire upwards at an oblique angle. At first the pilots were dubious of the scheme, but on the night 17/18 August, *Gefr* Hölker of 5./NJG 5 shot down two bombers during the Peenemünde raid. He was followed by *Lt* Erhardt of 6./NJG 5 who shot down four bombers within 30 minutes.

From July 1943 daylight attacks on the German homeland by USAAF heavy bombers began to increase in intensity. By the end of August no less than twelve fighter *Gruppen* had been transferred to the defence of the Reich. These included

Three Bf 110G–2 fighter-bombers in flight over the sea. It was quite usual to see Luftwaffe *aircraft of the same period sporting different camouflage patterns.*

I. and II./ZG 1 and III./ZG 26, the former *Gruppe* later becoming I./ZG 26 as part of a new *Geschwader* under *Maj* Boehm-Tettelbach. By the end of October a new ZG 76 had been established under *Oberstlt* Rossiwall, all units apart from II./ZG 26 being equipped with the Bf 110.

The early raids by USAAF bombers met with heavy opposition both from the single-engined fighters and the Bf 110 and Me 410 destroyers of the *Luftwaffe*. Early in January 1944 a Group of Mustang fighters, able to stay with the bombers throughout their sortie, joined a USAAF raid. From then on the *Luftwaffe* ceased to have it all its own way, and losses to German fighters, particularly the rocket-firing Bf 110s, rose alarmingly.

Perhaps the most disastrous operation was that on 16 March 1944 when 43 Bf 110s from ZG 76 attacked a USAAF force attempting to bomb Augsburg. At first the Messerschmitts attacked successfully, but then the escort pounced. Unable to survive in combat with Hurricanes and Spitfires of 1940 vintage, the Bf 110s were cut to pieces by Mustangs and Thunderbolts. No less than 26 Bf 110s were shot down, approximately 60 per cent of the total force.

By June 1944 the *Zerstörergruppen*, with the exception of II./ZG 1 under *Hptm* Albrecht, had been re-equipped with the Me 410. One of the last operations by this unit came on 2 July when it intercepted a USAAF formation over Budapest. Shortly afterwards it was redesignated III./JG 76 and re-equipped with the Bf 109G.

During the light summer nights of 1943, the visual search tactics employed by the *Nachtjagdgruppen* had proved very successful, but they were no more than a stop-gap pending the introduction of new equipment to operate in spite of 'Window'. At the end of 1943 three new bomber detection devices entered service with the German night fighter force: *Lichtenstein* SN–2, *Naxos* and *Flensburg*. The SN–2 was a new night fighter radar which was not jammed by the 'Window' then in use by the RAF. *Flensburg* and *Naxos* were radar homing devices, the former operating on the emissions of the British Monica tail-warning system and the latter on the emissions of the H2S blind bombing device.

By January 1944 the *Luftwaffe*'s night fighter force had reached its zenith. Six complete *Geschwader* were operational; NJG 1 under *Oberstlt* Streib, NJG 2 under *Maj* Prinz zu Sayn-Wittgenstein, NJG 3 under *Oberstlt* Lent, NJG 4 under *Oberstlt* Thimmig, NJG 5 under *Oberstlt* Radusch and NJG 6 under *Maj* Schaffer. In addition the so-called '*Eisenbahn-Nachtjagd*' units I. and II./NJG 100 and I./NJG 200 were operational in Russia. These *Gruppen* were quickly transferable, having their ground control based in railway carriages.

Although the Bf 110 still provided the

Right: *The Bf 110G–2/R1 which carried the huge 37 mm Flak 18 gun beneath the fuselage. The square object beneath the gun mounting is a shell ejector chute.*

Nachtjagdverbänd with 60 per cent of its equipment, a gradual switch was being made to the Ju 88, Do 217 and He 219.

The new equipment then being delivered to the *Luftwaffe*'s night fighting force enabled it to inflict heavy losses on the RAF during the early months of 1944. On 21 January, 55 bombers were lost out of a force of 648 attacking Magdeburg and a week later 43 were shot down out of 683 raiding Berlin. On 19 February, 78 bombers were shot down from a force of 823 bombing Leipzig and on 24th March, 72 were destroyed out of 823 attacking Berlin. The greatest night battle, indeed the greatest air battle of all time came on 30 March. The force of 795 bombers attacking Nuremberg was successfully engaged by 20 *Gruppen* of night fighters, about 200 aircraft. In the slaughter that followed, 94 bombers were shot down.

Between 18 November 1943 and 31March 1944 RAF Bomber Command lost 1,047 aircraft, about three-quarters of them to night fighters. Several pilots amassed extremely high scores during this period. By the autumn of 1944 *Oberstlt* Lent had gained 102 victories, but was fatally wounded on 5 October when his Ju 88 crashed into high tension wires near Paderborn.

Some months previously, on 21 January 1944, two famous night fighter pilots had been killed; *Maj* Prinz zu Sayn-Wittgenstein was shot down by a Mosquito after destroying five RAF bombers and *Maj* Meurer was killed when his He 219 collided with a bomber near Magdeburg. On 12 March the other 'fighting prince' *Maj* Prinz zur Lippe-Weissenfeld was killed when his Bf 110 crashed in the Ardennes.

Top: *The Bf 110G–4/R1 night fighter with an early type of radar aerial array.*

Above: *The more advanced array carried by the FuG 220* Lichtenstein *SN–2 equipped Bf 110G–4d/R3.*

Before the end of the war, Lent's score was topped by another Bf 110 pilot, *Maj* Heinz Schnaufer. Schnaufer ended the war as *Kommodore* of NJG 4 with 121 victories. His most successful 24 hours came on 21 February 1945 when he shot down two bombers early in the morning, and that evening he managed to work his Bf 110 into a bomber stream and down seven more. Schnaufer survived the war only to be killed in a road accident in France on 13 July 1950.

From the invasion of France until the end of the war in Europe, the story of the German night fighter force was one of unremitting decline. First the capture of France punched a great hole in the air reporting network, seriously reducing the warning time available to the defenders. Then the RAF started jamming the SN–2 radar, rendering it useless. From the autumn of 1944 the fuel reserves began to waste away, as one by one the refineries and storage centres were bombed to ashes.

On 31 December 1944 the German night fighter force had 913 aircraft on strength, but only 150 of these were Bf 110s. Many aircraft were grounded due to lack of fuel and during this period many were called upon to operate in the ground-attack rôle. The Bf 110s carried a rack for two 550 lb bombs under the fuselage and, if no drop tanks were fitted, two 110 lb bombs under each wing. Night ground attack was a particularly hazardous business, and losses were accordingly high.

Also during the final months of the war, the German night fighter crews were the subject of the unwelcome attentions of the Mosquitos of No. 100 Group, RAF. On 5 March 1945, *Oberstlt* Borchers, *Kommodore* of NJG 5 was shot down by a Mosquito flown by Wing Commander Gibb and several other night fighter aces were lost in this way.

Just over a month later came the collapse of the Third Reich. So it was that the Bf 110 ended its career, in action from the first day of the war to the last; and the two month's thrashing it took during the Battle of Britain clouded the valuable work that it performed for five years in the defence of Germany. It is true that the Bf 110 had to be relegated to night fighting because it could not survive in the daytime, but this was true of many famous aircraft, not least of which was the Lancaster, the Bf 110's main opponent.

Specification (Bf 110G–4a)

SPAN	16·20 m	(53 ft 4¾ ins)
LENGTH	12·65 m	(41 ft 6¾ ins)
HEIGHT	4·00 m	(13 ft 1½ ins)
EMPTY WEIGHT	5,100 kg	(11,245 lbs)
LOADED WEIGHT	9,400 kg	(20,727 lbs)
MAXIMUM SPEED	550 km/h	(342 mph)
MAXIMUM RANGE	2,100 km	(1,305 miles)

The Bf 110G–4b/R3 which carried both FuG 212 and FuG 220b radar equipment. All night fighters had large flame-damping exhaust systems and were usually fitted with auxiliary drop tanks.

The uncompleted Bv 155 V3 which was shelved in favour of the V4. The Bv 155 was developed from an original Messerschmitt design.

Me/Bv 155

THE MESSERSCHMITT Me 155 project qualifies for inclusion in this history mainly because it was eventually built by the Blohm und Voss company of Hamburg. The original Me 155 was projected early in 1942 as a shipboard fighter for use aboard the aircraft carrier *Graf Zeppelin*. It was essentially similar to the Bf 109G but featured longer undercarriage legs which retracted inwards into the wings, catapult spools and an arrester hook. Powered by a 1,475 hp Daimler-Benz DB 605A engine, it was anticipated that the Me 155 would attain a maximum speed of 403 mph.

Shortly afterwards the RLM lost interest in shipboard fighters so that in November 1942 the design was reworked as the Me 155A fighter-bomber. Essentially similar to the previous project, the Me 155A had its carrier equipment removed and replaced by a rack for a 2,200 lb SC 1000 bomb. To enable this weapon to clear the ground the retractable tailwheel was replaced by a fixed unit taken from a Bf 110. The Me 155A carried an armament of three 20 mm MG 151/20 cannons and was intended to achieve a speed of 404 mph without its bomb load.

After the project again failed to win a production contract it was redesigned as a high-altitude fighter. The original project, designated Me 155B, was essentially similar to the Me 155A apart from the installation of a 1,500 hp DB 628 engine with a ducted spinner. This design was replaced by another which featured a new square wing centre section, the tail of the Me 209 V5 and a DB 603A engine with TKL turbo-supercharger.

Construction of the Me 155B was about to begin when, in August 1943, the RLM ordered that the whole project be handed over to Blohm und Voss because of the latter's spare design capacity. The chief designer of Blohm und Voss, Dr Vogt, immediately proposed a complete redesign of the project which did nothing to cement relations between the two companies! Eventually the Bv 155 V1 emerged on 8 February 1945 still retaining some Bf 109 components, but differing considerably from Messerschmitt's original proposal. The Bv 155 V2 and V3 were similar, intended as forerunners of the Bv 155B-1 production model, but the latter aircraft was shelved in favour of the Bv 155 V4. This model, prototype for the Bv 155C-0, was even more sophisticated, but neither it, nor the Bv 155 V5, V6 or V7 were completed.

Specification (Me 155A—estimated)

SPAN	9·92 m	(32 ft $6\frac{1}{2}$ ins)
LENGTH	9·02 m	(29 ft 7 ins)
HEIGHT	3·40 m	(11 ft 2 ins)
LOADED WEIGHT	4,500 kg	(9,932 lbs)
MAXIMUM SPEED	(650 km/h (clean) and 510 km/h with bomb)	(404 mph (clean) and 317 mph with bomb)

Bf 161

Parallel with the development of the Bf 162 *Jaguar* high speed bomber, Messerschmitt began work on a similar fast reconnaissance aircraft to be known as the Bf 161. As no official requirement existed for such a machine, the Bf 161 was allocated a much lower development priority than the *Jaguar*. The first prototype did not fly for almost a year after the maiden flight of the Bf 162 and only two aircraft were completed.

The first of these was the Bf 161 V1 (D–AABA) which flew in the spring of 1938. It was powered by two 960 hp Daimler-Benz DB 600Aa twelve cylinder in-line engines and was, like the Bf 162, based on the Bf 110 destroyer. It differed mainly in carrying a crew of two and having a longer, deeper forward fuselage, the nose of which was extensively glazed to accommodate the observer. Defensive armament was reduced to one forward firing 7·9 mm MG 15 machine gun, and provision was made for a hand-held camera.

In the autumn of 1938 the Bf 161 V2 joined the test programme, this differing from the V1 in having a projecting upper portion to the glazed nose. The aircraft was later used to test the remotely-controlled gun barbettes eventually fitted to the Me 210 and Ar 240, but no production was undertaken.

Specification
No reliable details available.

Only two prototypes of the Bf 161 reconnaissance aircraft were completed, the V2 differing in featuring a projecting upper nose section.

Bf 162 Jaguar

The public revelation of the *Luftwaffe* to the world on 1 March 1935 led to several specifications being issued to German aircraft manufacturing firms for new types. One of the most important of these was that for a twin-engined fast bomber placed with the Focke Wulf, Henschel, Junkers and Messerschmitt companies. Focke Wulf, after the failure of their Fw 57 design, declined to participate, but the other companies produced respectively the Hs 127, the Ju 88 and the Bf 162.

Messerschmitt's design, unofficially named *Jaguar*, was basically similar to the Bf 110 destroyer, but featured a new wider forward fuselage with an abbreviated cockpit canopy and a glazed nose. Carrying a crew of three, the aircraft was powered by a pair of Daimler-Benz DB 600A liquid cooled in-line engines which produced 960 hp for take-off. Offensive loads could comprise two 550 lb SC 250 bombs mounted on external wing racks and a number of 110 lb SC 50 bombs in a small internal bay.

The first prototype, the Bf 162 V1 (D–AIXA), flew during the spring of 1937, being followed in September by the Bf 162 V2 (D–AOBE). Both aircraft carried a defensive armament of two 7·9 mm MG 15 machine guns on flexible mountings, one in the nose and one in the rear of the cockpit. On 20 August 1938 the third and final prototype, the Bf 162 V3 (D–AOVI), made its first flight. However, by then the Ju 88 had been chosen for production and the Hs 127 and Bf 162 were abandoned.

During the early war years the three Messerschmitt *Jaguar* prototypes were used for development work, the Bf 162 V3 serving as a test bed for the FDL 131 remotely-controlled gun barbettes which were installed in the Me 210. The RLM type number 8–162 was later re-used by the Heinkel He 162 *Salamander* jet fighter.

Specification (Bf 162 V2)

SPAN	17·16 m	(56 ft $3\frac{1}{2}$ ins)
LENGTH	12·75 m	(41 ft 10 ins)
HEIGHT	3·58 m	(11 ft 9 ins)
MAXIMUM SPEED	480 km/h	(298 mph)

Bf 163

Not to be confused with the Me 163 *Komet*, the Messerschmitt Bf 163 was a STOL reconnaissance aircraft designed to compete with the Fieseler Fi 156 *Storch* and the Siebel Si 201. It was designed by *Dipl Ing* Robert Lüsser under the project designation P.1051 and was a high-wing monoplane with automatic leading edge slots, double slotted flaps and a very tall undercarriage. The most unusual feature of the design was its variable incidence

The Bf 162 V2 Jaguar *(D–AOBE) in flight. This fast bomber based on the Bf 110 was abandoned in favour of the Junkers Ju 88.*

wing which could be tilted at the main spar to an angle of 25 degrees from the horizontal. The aircraft was to be powered by a 240 hp Argus As 10C engine.

Production of the Bf 163 was to be undertaken by the Weser company because of Messerschmitt's pre-occupation with the Bf 109. An order was placed for three prototypes and ten pre-production Bf 163A–0 aircraft, but this was cut back to two in October 1937. The Bf 163 V1, D–IUCY (w/nr 822), flew for the first time on 19 February 1938, but the second aircraft was never completed. Although the RLM preferred the Bf 163 to the Si 201, both aircraft were abandoned in favour of the cheaper Fi 156. No photographs of the Bf 163 V1 have survived.

Specification (Bf 163 V1)

SPAN	13·27 m	(44 ft 6½ ins)
LENGTH	9·73 m	(31 ft 11¾ ins)
EMPTY WEIGHT	995 kg	(2,193 lbs)
LOADED WEIGHT	1,310 kg	(2,888 lbs)

No performance figures available.

An unusual front view of the Me 163 V1 after being fitted with the Walter R II–203 rocket engine.

Me 163 Komet

MANY UNUSUAL combat aircraft were developed during the Second World War, but perhaps the most startling was Germany's Messerschmitt Me 163 *Komet*. The aircraft can still claim to be the only fighter powered solely by a rocket engine to see operational employment. It was also notable in being the first tailless aircraft to enter squadron service.

Although possessing a maximum speed in excess of any other operational aircraft of its time, the *Komet* also suffered from many drawbacks. These included its extremely short range, its tendency to blow up at the slightest provocation and the corrosiveness of its fuels which often resulted in severe burns to the pilot.

The Messerschmitt 163 was designed by Dr Alexander Lippisch who, as early as 1926, had built a tailless glider known as the *Storch*. This glider was later fitted with an 8 hp engine and made its first powered flight on 17 September 1929. Besides developing tailless aircraft, Lippisch, who was at this time working for the Rhön-Rossitten *Gesellschaft*, also initiated the design of the delta wing. His first machine to employ this layout was the *Delta* I which flew in the summer of 1930. Although failing to gain a C of A, the machine was followed two years later by the more successful *Delta* III built by the Focke-Wulf company.

The first Lippisch design to achieve prominence was the canard *Delta* IV which was also completed in 1932 by the Fieseler company as the F–3 *Wespe*. Powered by two 75 hp Pobjoy R engines, the two-seat machine crashed during its maiden flight. Fieseler immediately lost interest in the *Delta* IV, but Lippisch eventually repaired the aircraft and removed the pusher engine and the foreplane. Now designated *Delta* IVa, the machine possessed vile handling characteristics and again crashed.

This crash led to the RLM placing a ban on the construction of further tailless aircraft, but this was lifted after Dr Walter Georgii of DFS intervened. DFS, which had grown out of Rhön-Rossitten *Gesellschaft*, took over Lippisch's design, developing it as the *Delta* IVb. The new design featured trailing edge sweep back to the wings together with anhedral tips. It was flown quite successfully and was followed in 1936 by the *Delta* IVc (D–ENFL). Designated DFS 39 by the RLM, this machine featured endplate fins in place of the anhedral wingtips, and after exhaustive testing by Heini Dittmar, received a full C of A.

As early as 11 June 1928, Fritz Stamer made the first rocket powered flight using a glider powered by solid fuel units. In September 1929 von Opel attempted to repeat the experiment, but his aircraft caught fire and crashed. It was not until the autumn of 1936 that the first flight by a liquid fuelled rocket engine was made when a simple 300 lb thrust Walter unit was tested by a Heinkel He 72 biplane.

The success of these experiments led the Heinkel company to donate two He 112 airframes for use in connection

The Me 163 V1 begins a 'sharp' start with its rocket engine belching forth smoke and steam.

with a rocket engine developed by Prof Wernher von Braun. During 1937 an He 112 was fitted with a variable thrust liquid fuel engine developed by Prof Hellmuth Walter. The 2,000 lb thrust unit proved so successful that an aircraft powered solely by rocket propulsion was built by Heinkel, the He 176.

The testing of the DFS 39 described earlier led the RLM to request the construction of a second aircraft, this to be powered by the 1,100 lb thrust Walter R I–203 rocket engine then under development. Because of limited production facilities at DFS, this company was to produce the wings only, the task of building the fuselage being transferred to the Heinkel company.

Work continued at a very slow rate, mainly because Lippisch was also involved in two further projects, the DFS 40 *Delta* V and DFS 194. Eventually, on 2 January 1939, the whole team of 20 scientists and engineers was transferred to Messerschmitt at Augsburg, work then being concentrated on the more promising DFS 194. The team, known as Section L, was transferred to Peenemünde-West early in 1940 and later in the year the DFS 194, piloted by Heini Dittmar, reached a speed of 342 mph with a down-rated Walter R I–203 engine.

Following the success of the DFS 194, a definitive variant was produced, designated Me 163. This aircraft had a small rotund fuselage with low aspect ratio

An aerial shot of the Me 163 V1 after its first powered take-off. The fantastic climbing ability of the Komet *had to be seen to be believed.*

swept wings, large vertical tail surfaces and no horizontal stabiliser. The first two prototypes, the Me 163 V1 and V2 were completed during the spring of 1941, initial flight trials being carried out without the rocket engine, the machine being towed aloft by a Bf 110 tug.

During the summer of 1941, the Me 163 V1 was taken to Peenemünde-Karlshagen and fitted with a 1,650 lbs thrust Walter R II–203 engine. It made its first powered flight in August 1941, again piloted by Heini Dittmar. A number of successful flights followed, and on 2 October, Dittmar cast off the machine after being towed up to 13,000 ft. A few moments later the aircraft attained the astonishing speed of 623·85 mph, over 150 mph faster than the official world speed record, then held by the Me 209 V1!

Although very promising, the R II–203 engine installed in the first prototypes burned very erratically and proved the cause of many accidents. Therefore a new engine, the HWK 509, was developed still using T-*Stoff* (hydrogen-peroxide and water) as fuel, but with C-*Stoff* (hydrazine hydrate and methyl alcohol) in place of Z-*Stoff* as the catalyst. The new 3,750 lbs thrust engine was installed in the third prototype, the Me 163 V3, construction of which began on 1 December 1941.

The third prototype was the forerunner of the proposed Me 163B operational aircraft. Apart from the new engine, the V3 differed in having an even bulkier fuselage and the wing leading edges with constant sweep-back incorporating fixed slots which made the aircraft virtually impossible to stall. The aircraft took-off with the aid of a wheeled dolly which was jettisoned at an altitude of between 15 and 30 feet. Landing was accomplished with the aid of a specially-sprung retractable skid.

During the early autumn of 1941, *Lt* Rudolf Opitz joined Dittmar as test pilots for the Me 163 programme. Several unpowered flights were made with the Me 163 V3 during the summer of 1941, but later in the year, Dittmar was forced to leave the programme when the skid of his rocket-powered Me 163A failed to extend, damaging his spinal column. After this accident the pilot's seat was specially sprung.

It was not only the skid that caused trouble with the revolutionary aircraft however. On several occasions the jettisonable dolly was released too soon and bounced back to strike the aircraft. But it was the power plant that caused most of the accidents. The inadvertant mixture of the tiniest quantities of fuel could cause an explosion, and care was taken to paint the T- and C-*Stoff* containers different colours. Before each run, the complete engine was swilled out with water under pressure to check for any leaks. But, despite these precautions, many aircraft exploded when attempting to take-off or land, leaving merely a large circle of tiny shattered fragments. The other main danger was with the highly corrosive nature

Left: *The Me 163 V3 was a development aircraft for the Me 163B. Initially unpowered, the prototype was the first to incorporate leading edge slots.*

Right: *A standard production Me 163B–1a (w/nr 191454) target defence fighter. The B–1a differed from earlier variants in having two 30 mm cannon in the wing roots.*

Above: *The Me 163 V8 was the production-prototype for the B-series.*

of T-*Stoff*, which in one case literally dissolved a pilot alive. A rubber suit was worn, but this only gave a slight protection from such fuel leaks.

The first powered flight was made with the Me 163 V3 in August 1943 and by this time well over half of the initial order for 70 aircraft had been completed. The pre-production aircraft all bore *Versuchs* numbers, the last recorded being the Me 163 V40. Apart from the B-series aircraft, a small batch of ten Me 163A unpowered gliders were built by Wolf Hirth for training purposes.

Early in 1943 a special test unit was established at Peenemünde-West under *Hptm* Wolfgang Späte who had joined the test programme in July 1942 from 5./JG 54. Designated *Erprobungskommando* 16, the unit initially comprised four pilots in addition to Späte and Opitz. These were *Hptm* Toni Thaler, instructor and ferry pilot, *Oblt* Josef Pöhs from II./JG 54, *Oblt* Johannes Kiel from I./ZG 26 and *Lt* Herbert Langer from JG 53.

Training commenced using ordinary gliders such as the Granau Baby and DFS Kranich, but following the bombing of Peenemünde in July 1943, EK 16 was forced to move to Bad Zwischenahn near Oldenburg. During the autumn the first of 30 selected pilots reached the unit, and training continued using *Habicht* gliders with successively shorter wing spans. The trainee was then switched to the unpowered Me 163A and finally made a 'sharp' start in the rocket-engined A-series *Komet*.

Later in 1943 the first 'powered egg' (as the Me 163B was nicknamed) was delivered to EK 16. The operational *Komets* were originally equipped with two

20 mm MG 151/20 cannons in the wing roots under the designation Me 163Ba–1. Because of the extremely short range of the Me 163 (approximately 50 miles) it was realised that the machine could only be used for target defence operations. Späte proposed the establishment of a string of bases right across Germany, approximately 90 miles apart, over which any attacking bombers would have to pass.

Just prior to the formation of the first *Komet* operation *Staffel*, Späte was unaccountably transferred to the eastern front to command IV./JG 54, his place at EK 16 being taken by Toni Thaler. The first operational squadron was formed in May 1944 under *Hptm* Robert Olejnik, a holder of the *Ritterkreuz*, who had earlier been transferred from III./JG 1 to EK 16. The unit was designated 1./JG 400 and was established at Wittmundhafen. A month or so later, 2./JG 400 was formed at Venlo in Holland under *Hptm* Böhner and by July 1944 the completed I.*Gruppe*/JG 400 had been transferred to Brandis near Leipzig. The task of the *Gruppe* was to defend the synthetic oil plant at Leuna, some 30 miles to the west.

The operational squadrons were now equipped with the Me 163B–1a which differed from the earlier model in being equipped with two 30 mm MK 108 cannons, a *Revi* 16B gunsight and full armour protection. Having no conventional undercarriage, the ground handling of the Me 163 proved somewhat of a problem until the '*Scheuschlepper*' tractor was developed. This was fitted with hydraulic arms which lifted the aircraft off the ground and allowed it to be transported back to its hangar.

One of Erprobungskommando *16's Me 163Bs undergoing a routine maintenance. During this all fuel lines to the rocket motor were flushed out with gallons of water.*

Hptm Olejnik was injured after his *Komet* suffered a flame-out and he was posted to the *Ergänzungstaffel* of JG 400 which was then being formed. Command of 1./JG 400 then passed to *Hptm* Fulda, 3 *Staffel* being led by *Hptm* Falderbaum, an ex *Wilde Sau* night fighter pilot. Operations from Brandis by I./JG 400 under *Hptm* Rudolf Opitz often proved somewhat abortive, mainly because of lack of servicing facilities. Fuel had to be stored in the railway wagons that brought it, the airfield itself was too short and the repair facilities were inadequate. The few *Komets* that were scrambled to intercept Allied bomber formations were hampered both by their short range and inadequate radio control network. Although some successful interceptions were made, it is unlikely that more than nine Allied aircraft were destroyed by the Me 163.

During October 1944 a *Geschwader Stab* of JG 400 was established under *Maj* Wolfgang Späte who had been recalled from IV./JG 54 to command the unit. A month or so earlier, the establishment of a second *Komet* equipped *Gruppe* was proposed at Stargard near Stettin. A special concrete runway was built some distance from the main airfield and a large new hangar was constructed. In November, the first '*Scheuschlepper*' tractors arrived followed a month later by the first Me 163s.

The unit became operational in December 1944 and designated II./JG 400, was placed under the command of *Hptm* Rudolf Opitz; *Hptm* Fulda taking over control of I.*Gruppe* which remained at

One of JG 400's Komets *which was captured after the war and transported to the United States for evaluation.*

Brandis. About four or five training flights per day were made by the *Gruppe*, maintenance on the aircraft being carried out by I./EJG 1 which was also based at Stargard. Ground handling equipment included five tractors and six special fuel tankers each with a capacity of 440 Imp gallons. The tankers were clearly marked T- and C-*Stoff* on the side to prevent any possible admixture of the two fuels.

As has been mentioned previously, *Hptm* (later *Maj*) Olejnik left 1./JG 400 to supervise the establishment of a training and replacement *Komet* squadron at Udetfeld. Originally known as the *Ergänzungstaffel*/JG 400 and later IV./EJG 2. The *Gruppe* comprised three *Staffeln*: 13./EJG 2 under *Oblt* Adolf Niemeyer, 14./EJG 2 under *Lt* Mano Ziegler and 15./EJG 2. Pupils were usually aged between 19 and 23 and joined the unit after a short course at a glider school.

The training course with IV./EJG 2 lasted about six weeks and was divided into seven distinct sections. Firstly the pupil made five gliding flights in the Me 163A. He was towed up to about 5,000 ft by a Bf 110 and released to make a gliding landing. During the second stage of the course, the pilot completed four or five powered flights in the Me 163A of about five minutes duration and one full duration powered flight. The fourth, fifth and sixth stages of training were similar but were carried out on the Me 163B. The final stages of training involved 12 to 18 hours of supervised general flying practice.

Apart from the difficulties previously detailed, it had proved extremely difficult to even make hits on an enemy aircraft. This was mainly due to the extremely fast closing speed of the Me 163 (about 700 ft/min) and the slow rate of fire of the MK 108 cannon, which were also prone to jamming. One Me 163A was fitted with 24 rockets, but although these were successful, it still proved virtually impossible to aim them accurately.

Perhaps the most advanced solution to the problem was that suggested by Dr Langweiler, the inventor of the *Panzerfaust* anti-tank weapon. He proposed the installation of ten vertically-mounted tubes in the wing roots of the *Komet*, each containing a 50 mm shell. The upward-firing device, known as the SG 500 *Jagdfaust*, was aimed by means of a photo-electric cell; the shadow of the bomber passing overhead automatically triggering the weapon. Tests were carried out with the SG 500 installed in both a Fw 190 and a Me 163 by *Lt* Hachtel of EK 16, and proved very successful. The target comprised a strip of fabric fixed beneath two barrage balloons, but after this was pronounced too conspicuous to Allied reconnaissance aircraft by the RLM, the test target was mounted between two high poles.

After the erection of the poles, Hachtel took-off to make the first test, but it ended in disaster when his *Komet* crashed and severely damaged his spine. Testing was taken over by *Lt* Fritz Kelb and twelve *Komets* were eventually fitted with the device. Although the SG 500 equipped aircraft never saw service, Kelb did destroy a B–17G Fortress with the weapon.

The temperamental nature of the Walter HWK 109–509 engine led to the placing of an order for the development of an alternative engine with the BMW company. Designated P.3390A, the engine used M-*Stoff* (methanol) and SV-*Stoff* (nitric acid) propellants and thrust was controllable from 660 lbs to 3,300 lbs. Eventually the engine, then redesignated BMW 109–510, was test flown by the Me 163 V10, but troubles with the fuel pumping system prevented successful employment before the end of the war. BMW also produced the 109–718 engine which was designed to give 5,500 lbs thrust using a mixture of SV- and R-*Stoff*.

Development of the Walter engine eventually resulted in the HWK 109–509C which featured an auxiliary cruising chamber producing an extra 660 lbs thrust. The engine increased powered flight endurance from eight to twelve minutes and was first tested in the Me 163 V6 early in 1944. The new unit was chosen for installation in the Me 163C which also introduced a pressurised cabin enclosed by a bubble canopy. Production prototype for the C-series was the Me 163 V18 which attained a speed of no less than 702 mph on 6 July 1944, severely damaging its rudder in the process. In the event, only three Me 163C–0 pre-production aircraft were completed before the end of the war, and the production model, the Me 163C–1a, failed to see operational service.

In March 1945 the *Geschwader Stab* of JG 400 was disbanded and Späte was transferred to command III./JG 7 equipped with the Me 262. I./JG 400 was also disbanded, many of its personnel spending the last weeks of the war fighting with the German army in Czechoslovakia. In April 1945 the one remaining *Komet* equipped *Gruppe*, II./JG 400, moved to Husum in Schleswig Holstein. Its one combat casualty occurred during the operation when one of its Bf 110 towplanes was shot down by an RAF Tempest fighter. After the war many of II./JG 400's Me 163B–1as which were captured by British forces, were transported to England for evaluation.

Thus ended the career of one of the most remarkable combat aircraft of the war, an aircraft which caused more casualties to friend than foe and demanded qualities of extreme heroism from all its pilots. It is perhaps some small tribute to this band of men that the British Air Ministry later forbade all powered flight trials with the *Komet* as they considered them too dangerous!

Specification (Me 163B–1a)

SPAN	9·32 m	(30 ft 7 ins)
LENGTH	5·69 m	(18 ft 8 ins)
HEIGHT	2·74 m	(9 ft 0 ins)
EMPTY WEIGHT	1,905 kg	(4,191 lbs)
LOADED WEIGHT	4,110 kg	(9,042 lbs)
MAXIMUM SPEED	960 km/h	(596 mph)
ENDURANCE	8 mins	(8 mins)

Left: *The Me 163 V6 was fitted with an auxiliary cruising chamber for tests in connection with the Me 163C programme.*

Right: *Despite being fitted with a nosewheel undercarriage, the Me 208 V1 was similar to the Bf 108B* Taifun.

Me 208

THE SUCCESS of the Bf 108 led Messerschmitt to develop an improved version under the designation Me 208. The new machine differed mainly in having a nose-wheel undercarriage, the 240 hp Argus As 10C air-cooled in-line engine of the Bf 108B being retained. Construction of two prototypes, designated Me 208 V1 and V2, was completed at the S.N.C.A. du Nord factory in France, but no production was undertaken.

After the end of the Second World War, the Me 208 was redesigned as the Nord 1100 *Noralpha* and large-scale production of a 240 hp Renault 6Q–10 powered version, the Nord 1101 *Ramier*, was undertaken. Eventually 200 *Ramiers* were built for use by both the French Air Forces and several civilian companies. An experimental development powered by a 240 Potez 6D.30 engine was also built, designated Nord 1104. Development of the basic Me 208 continued until the early 1950s, but these more advanced variants bore little relation to the original German design.

Specification (Me 208 V1)

SPAN	11·50 m	(37 ft 8¾ ins)
LENGTH	8·85 m	(29 ft 0½ ins)
HEIGHT	3·35 m	(11 ft 0 ins)
EMPTY WEIGHT	945 kg	(2,084 lbs)
LOADED WEIGHT	1,580 kg	(3,484 lbs)
MAXIMUM SPEED	305 km/h	(190 mph)
NORMAL RANGE	1,200 km	(746 miles)

Me 209 V

FOR JUST OVER thirty years the world speed record for piston engined aircraft was held by the Me 209 V1 until finally, on 16 August 1969, the record was beaten by a Grumman Bearcat.

The Me 209 was in fact conceived as the successor of the Bf 109, being the first Messerschmitt design to bear the 'Me' designation. The design was initially proposed as Projekt 1059; a high-speed fighter to be powered by a specially modified DB 601 engine. The Me 209 was more or less a flying engine with the cockpit positioned near the rear of the fuselage with tiny fin and rudder surfaces and a short span wing.

Messerschmitt's chief Augsburg test

For over thirty years the world speed record for piston-engined aircraft was held by the Me 209 V1.

pilot, Dr Hermann Wurster first flew the Me 209 V1 (D–INJR—w/nr 1185) on 1 August 1938, the machine being powered by a standard 1,100 hp DB 601A engine. The first flight lasted only seven minutes and subsequent tests resulted in no less than 17 major complaints being listed about the aircraft. Perhaps most important of these were instability both on the ground and in the air, a poor view from the cockpit, and very high stick forces.

The second aircraft (D–IWAH—w/nr 1186) made its initial flight on 8 February 1939 with *Flugkapitän* Fritz Wendel at the controls. Wendel assumed the rôle of chief test pilot after Wurster left to concentrate on the engineering side of the company. Two months after its first flight, the Me 209 V2's engine seized and it crashed. The Me 209 V3 (D–IVFP—w/nr 1187) was intended for an attempt on the world speed record, but after the Heinkel He 100 V8 had smashed it with a speed of 463·92 mph in March, it was decided to re-engine the Me 209 V1 for a Messerschmitt attempt on the record.

A specially boosted Daimler-Benz DB 601ARJ engine was fitted to the Me 209 V1, the unit being capable of producing the enormous power of 2,300 hp for a short period. The aircraft was painted dark blue overall and with Fritz Wendel at the controls, broke the record with a speed of 469·22 mph on 26 April 1939. Heinkel refused to accept that the Me 209 was superior to his He 100 and he planned another attempt on the record which was only finally prevented by the outbreak of war. The Me 209 V1 was, incidentally, publicised by the *Nazi* propaganda machine as the 'Me 109R' in an attempt to boost the *Luftwaffe*'s major fighter.

The Me 209 V3 did eventually fly at the end of May 1939 and was followed by the V4, the only aircraft to be built as a pure fighter aircraft. The fourth machine (D–IRND—w/nr 1188) made its first flight on 12 May 1939, differing in having increased wing span with leading edge slots in an attempt to cure the vicious stalling characteristics displayed by the earlier prototypes. Armament was carried in the shape of a single 30 mm MK 108 cannon firing through the spinner and two MG 17 machine guns mounted above the engine cowling. An attempt to install two further guns in the wings was abandoned because of the thin airfoil section.

Despite all these modifications, the Me 209 V4 still proved a difficult machine to fly, Wendel himself describing it as 'a diminutive monstrosity'. Further attempts were made to improve the flight characteristics by again increasing the wing span and replacing the slots by drooped leading edges. The size of the radiators was also increased but all these modifications so affected performance that the Me 209 was little better than the standard Bf 109. This, coupled with its still uncomfortable handling characteristics, resulted in the final abandonment of the whole project.

Specification (Me 209 V4)

SPAN	10·04 m	(32 ft 11$\frac{1}{4}$ ins)
LENGTH	7·24 m	(23 ft 9 ins)
LOADED WEIGHT	2,800 kg	(6,174 lbs)
MAXIMUM SPEED	600 km/h	(373 mph)
SERVICE CEILING	11,000 m	(36,080 ft)

Fritz Wendel, Messerschmitt's chief test pilot, described the Me 209 V4 fighter as a 'brute'.

Me 209A

UNDISMAYED by the failure of their Me 209 V and Me 309 designs, Messerschmitt began development of yet another Bf 109 replacement—the Me 209A. Apart from retaining the RLM type number 209, the new design bore no relation to the record-breaking Me 209 V1 of 1939.

The first prototype of the new fighter was designated Me 209 V5, the four previous *Versuchs* numbers having been allocated to the earlier Me 209. Like the Me 309, the Me 209A featured an inwards retracting undercarriage, and was powered initially by a 1,750 hp DB 603A engine in an annular cowling. The installation of this engine necessitated an increase in wing area and the provision of a much taller tail fin and rudder assembly.

Flugkapitän Fritz Wendel took the Me 209 V5 up on its first flight on 3 November 1943. The machine, which was in fact an interim prototype for the proposed Me 209A–1 production model, was rebuilt from a Bf 109G–5 airframe. After only a few test flights the Me 209 V5 was fitted with the more powerful DB 603G engine which developed 1,900 hp for take-off. Trials proved that the Me 209 had excellent flight characteristics, although it was anticipated that the new thin section wing proposed for the production model would even further improve matters.

Because of the shortage of DB 603 engines, the RLM specified that an alternative model be developed to utilise the 1,750 hp Jumo 213E power plant. This variant was to be designated Me 209A–2, the DB 603 model being known as the A–1. Five production variants were proposed; the Me 209A–2/R2U1 with an armament of one 30 mm MK 108 and two 20 mm MG 151 cannon, the R2U2 with two extra MK 108s or MG 151s, the R2U3 and R2U4 which were similar but employed GM–1 power boosting, and the Me 209A–2/R2U5 which carried a 300 litre drop tank. In the event only one prototype of the A–2 was completed, designated Me 209 V6, the aircraft making its first flight in May 1944.

Much earlier, on 23 April 1943, Messerschmitt had proposed a high-altitude version of the Me 209 to be powered by the specially adapted DB 628 engine. The project was given the designation P.1091a and it was anticipated that the type would attain an altitude of 17,000 m (55,775 ft). The DB 628 engine was flight tested by the Bf 109 V49 and V50 and the variant finally emerged as the Me 209H.

The Me 209H V1 first flew in June 1944, powered by a 1,900 hp DB 603G engine. It had originally been intended to utilise a 2,000 hp DB 627 high-altitude engine driving a four-blade airscrew, but this unit was not ready in time. The Me 209H also featured an increase in wing span with radiators in the leading edge of the centre section. Despite the success of the aircraft, no production was undertaken as it was

Left: *As far as is known, this is the only surviving photo of the Me 209 V5, forerunner of the proposed Me 209A.*

felt that the Focke-Wulf Ta 152H would involve far less disruption of Germany's fighter production programme.

Specification (Me 209 V5)

SPAN	10·95 m	(35 ft 11 ins)
LENGTH	9·62 m	(31 ft 6½ ins)
HEIGHT	3·55 m	(11 ft 8¼ ins)
EMPTY WEIGHT	3,290 kg	(7,255 lbs)
LOADED WEIGHT	3,870 kg	(8,535 lbs)
MAXIMUM SPEED	725 km/h	(449 mph)
SERVICE CEILING	12,400 m	(40,683 ft)

Me 210

'THIS MONTH once again we have received no Me 210s for battle tests, so that the employment of the model has to be postponed for another month. I also have the impression that since the first prototype was built you have made too many changes. The prototype and the proposed machines in the series are so different that the results of testing are no use for the practical question of behaviour in action.

The first prototype Me 210 bore considerable similarity to the Bf 110. The aircraft carries the civil registration D–AABF.

'One thing, dear Messerschmitt, must be made clear between us, and that is that there must be no more losses of machines in normal ground landings as the rsult of a faulty undercarriage; this can hardly be described as a technical novelty in aircraft construction. All these unnecessary scandals and this waste of time call for higher standards in the testing of your new aircraft, and I shall report in that sense to my department.'

This letter, written to Messerschmitt by Ernst Udet, head of the RLM's Technical Department on 25 July 1941 describes in a few words the company's main failure, the Me 210. The aircraft was conceived in 1937 as a replacement for the Bf 110, able to undertake a much wider range of duties. As originally projected, the Me 210 was similar in appearance to the Bf 110, but featured a new and much more streamlined nose, with the cabin transparency placed further forward.

The prototype, the Me 210 V1, D–AABF (w/nr 0001), made its first flight on 2 September 1939. It was powered by two 1,100 hp Daimler-Benz 601Aa engines and featured a twin fin and rudder assembly. Even at this stage, the aircraft displayed marked longitudinal instability and was returned to the assembly shop to be fitted with a large single fin and rudder unit in an attempt to cure the trouble. The Me 210 V2, WL–ABEO (w/nr 0002), initially had the twin fin and rudder assembly, but was later similarly modified. The aircraft eventually crashed during flutter trials on 5 September 1940.

Even before the first flight of the Me 210 V1 the RLM had taken the unusual step of ordering 1,000 aircraft off the drawing board. In addition to the standard forward firing armament, the Me 210 V2 was also fitted with mock-ups of the FDL 131 gun barbettes one on either side of the fuselage aft of the cockpit canopy. These were remotely controlled barbettes, each containing a 13 mm MG 131 machine gun for rearward defence. The installation was first tested by the Bf 110C–2/U1 and was also later fitted to the Arado Ar 240 twin-engined fighter.

The Me 210 V3 was used for static tests and the V4 (w/nr 0004) was fitted with dive brakes and used for diving trials. Before the end of 1940, thirteen Me 210 prototypes had been delivered, all featuring various aerodynamic improvements. The Me 210 V14, V15 and V16 were delivered early in 1941 and the A–0 pre-production model began to leave the assembly lines during the spring of 1941. Although improved the aircraft still manifested severe longitudinal instability and had the

Left: *The Me 210 V2, WL–ABEO (w/nr 0002) in its final form was fitted with a large single fin and rudder assembly.*

An early production Me 210A–1 widely used in German propaganda shots of the period.

tendency to spin at the slightest provocation.

The first production model was the Me 210A–1 heavy fighter which carried an armament of two 20 mm MG 151/20 cannon and two 7·9 mm MG 17 machine guns firing forward and two MG 131 machine guns in the FDL barbettes. The Me 210A–2 fighter-bomber could carry two 1,100 lb bombs in an internal bay plus two 1,100 lb or 550 lb bombs on underwing racks.

Exceptionally, two 2,200 lb bombs could be mounted on the wing racks. The Me 210B–1 was a reconnaissance aircraft with the MG 17 machine guns removed to make way for two cameras, but only two aircraft were built. Three other variants were projected, the Me 210C–1, C–2 and D–1, respectively 1,475 hp DB 605 powered versions of the A–1, A–2 and B–1. The Me 210C–1 was built under licence by the Hungarian Manfred Weiss factory.

Because of the numerous modifications being introduced to the Me 210 on the assembly line, only 94 machines were delivered in 1941. The aircraft continued to suffer from severe longitudinal instability and finally, on 14 April 1942, production was ceased altogether. By this time

In this view of the Me 210A–1, the FDL 131 gun turrets are clearly visible.

some 200 machines had been completed, another 370 were on the assembly line and preparations were on hand for the production of a further 800.

After the suspension of work on the Me 210, another serious attempt was made to improve its flight characteristics. The fuselage length was increased by 3 ft $6\frac{3}{4}$ ins and leading edge slots were fitted which considerably reduced the aircraft's tendency to spin. These modifications were tested by the Me 210 V17 (which first flew with them on 14 March 1942). Production was then restarted, 95 machines being completed in 1942, 89 in 1943 and 74 in 1944.

During the winter of 1941–42, the Me 210 was delivered to II./ZG 1 on the Russian front, but by May the unit had returned to its Bf 110s. Following the introduction of the improved Me 210, a special experimental squadron was formed to introduce the machine into service in the west. Known as the *Versuchstaffel* 210, the unit was based at Soesterburg in Holland, being redesignated as 16./KG 6 in late August 1942. The aircraft made its first appearance over the British Isles on 6 September when two aircraft were shot down by high flying Typhoons over Yorkshire. After two more aircraft had been lost, Göring was heard to remark that his epitaph would read 'He would have lived longer if the Me 210 had not been produced'.

Late in August 1942, Me 210s were delivered to III./ZG 1 under *Maj* Roland Bohrt at Trapani in Sicily, the unit later moving to Gerbini. Desultory operations continued until May 1943, when the unit received the Me 410. The only reconnaissance unit to receive the Me 210B–1 was 2.(F)/122 based at Trapani, which operated the machine in company with its Ju 88s. By May, this unit too had received the Me 410.

Apart from German production, a few Me 210s were built by the Hungarian Manfred Weiss company for use by that Air Force's 102 Fast Bomber Group which eventually comprised three Squadrons.

The failure of the Me 210 led to the loss of over 600 aircraft from the German war effort, and the loss of over 30 million RM to the company. It also led Erhard Milch, who succeeded Udet after the latter's suicide, to demand the resignation of Willi Messerschmitt from the managing directorship of the Messerschmitt AG.

Specification (Me 210A–1)

SPAN	16·40 m	(53 ft $7\frac{3}{4}$ ins)
LENGTH	11·20 m	(36 ft $8\frac{1}{4}$ ins)
HEIGHT	4·30 m	(14 ft $0\frac{1}{2}$ ins)
LOADED WEIGHT	8,100 kg	(17,857 lbs)
MAXIMUM SPEED	620 km/h	(385 mph)
NORMAL RANGE	2,400 km	(1,491 miles)
SERVICE CEILING	7,000 m	(22,967 ft)

Me 261 Adolfine

THE PART PLAYED by Messerschmitt aircraft during the 1936 Olympic Games has previously been described: the aerobatics performed by Willi Stör's M 35, the success of the Bf 108 and the first public appearance of the Bf 109. But, even as the Games ended, Messerschmitt proposed the construction of an ultra-long-range aircraft to fly the Olympic flame non stop from Berlin to Tokyo where the 1940 contest was to be held.

The P.1064 as the project was designated, immediately captured the imagination of Hitler who allowed it to be named *Adolfine*. In appearance the Me 261, as it was dubbed by the RLM, was more or less a scaled-up Bf 110. It was powered by two 2,660 hp Daimler-Benz 606A/B engines which comprised two DB 601 units coupled together. A crew of five was carried including two pilots, a navigator, radio operator and flight engineer. Provision was made for personnel to rest in bunks during sustained flights.

Work on the Me 261 V1 began in the spring of 1939, but owing to the low development priority, it was not until 23 December 1940 that the machine was successfully flown. It was piloted by *Flugkapitän* Karl Baur, Messerschmitt's deputy chief test pilot and was followed, in the spring of 1941, by the second prototype.

It was proposed that the long range of the Me 261 might suit it for the task of maritime reconnaissance far out over the oceans. Consequently, the RLM requested that defensive armament be fitted to the second prototype. This task eventually proved impossible because the addition of gun positions would have reduced the fuel capacity of the aircraft to a point where it would have little range advantage over its competitors.

Early in 1943, the Me 261 V3 joined the test programme. It differed in having provision for a crew of seven and two 2,950 hp DB 610A/B engines. These comprised two 1,475 hp DB 605 twelve cylinder engines coupled together. In 16 April 1943, Baur flew a distance of 2,796 miles, staying in the air for no less than ten hours. The achievement was somewhat marred when the Me 261 V3's port undercarriage leg collapsed on landing.

After being repaired, the Me 261 V3 was delivered to the reconnaissance group of the high command of the *Luftwaffe* (Aufkl.Gr.Ob.d.L.) at Oranienburg. This specialised unit under the direction of *Oberst* Theodore Rowehl undertook many long-range missions using the Me 261 V3. These first two prototypes were badly damaged during an Allied attack on Lechfeld in 1944 and were scrapped.

Specification (Me 261 V3)

SPAN	26·87 m	(88 ft 1¾ ins)
LENGTH	16·69 m	(54 ft 9¼ ins)
HEIGHT	4·72 m	(15 ft 5¾ ins)
MAXIMUM SPEED	620 km/h	(385 mph)
NORMAL RANGE	11,000 km	(6,831 miles)

Left: *The Me 261 V2* Adolfine *at Lechfeld where it was severely damaged in 1944.*

Me 262

ALTHOUGH several remarkable combat aircraft were developed by the Messerschmitt company, it was perhaps the Me 262 that had the greatest single impact on the conduct of aerial warfare. The Me 262 was the first turbojet-driven fighter in the world to see operational service, although it only preceded the Gloster Meteor into action by a matter of days. The Me 262 was however the only jet to destroy enemy piloted aircraft in combat during the Second World War.

Serious jet engine development began in Germany in March 1936 when Ernst Heinkel afforded two young scientists, Pabst von Ohain and Max Hahn, full facilities to continue their pioneering work. By March 1938 the first successful engine, the HeS 3, was bench running, and experience eventually led to the world's first pure jet flight by the He 178 in August 1939.

Towards the end of 1938 work began to set up an official jet engine development group within the RLM's technical department. Heinkel was already sponsoring the design of two engine types, the 1,540 lbs thrust HeS 001 and the 1,760 lbs thrust HeS 006, and following approaches from the RLM, Bramo, BMW and Junkers set up their own programmes. After BMW took over Bramo in June 1939, they abandoned the complex 002 contra-rotating engine and concentrated on the more conventional 003 of their own design. Junkers proceeded with the conventional Jumo 004 unit which it was hoped would rapidly attain mass production status. Daimler-Benz showed little interest in the programme although they did begin work on the DB 007 project.

The development of these radically new power plants prompted the RLM to approach the Messerschmitt company with a view to the construction of a jet fighter. The research team, led by Woldemar Voigt, studied both single- and twin-engined layouts, and finally the latter was chosen, emerging as the Messerschmitt P.1065 project. Initially the turbojets were to be mounted in the centre of the gently swept-back wings, but because of difficulties associated with the positioning of the main spars, it was later decided that they should be underslung.

A wooden mock-up of the P.1065 project was completed and inspected by RLM officials on 19 December 1939. After the changes in turbojet position mentioned previously had been carried out, the RLM placed a development contract for three prototypes on 1 March 1940. The designation Me 262 was allocated and by early 1941, the three airframes had been completed.

Right: *An attempt was made to fly the Me 262 V1 with two BMW 003 turbojets in November 1941, but the turbine blades failed. This photo clearly shows the two-bladed airscrew in the nose.*

The second Me 262 first flew on 2 October 1942 powered by two Jumo 004 turbojets.

Although three Me 262s were now assembled, it was obvious that no turbojets would be ready to power them for at least six months. The BMW 003 was not flight tested until the summer of 1941 (beneath a Bf 110) and then only produced 990 lbs thrust. The Jumo 004 was bench run during November 1940, but it was not until March 1942 that the first test flight was made, again using a Bf 110. In order to test the aerodynamic characteristics of the new aircraft, the first prototype, the Me 262 V1, was fitted with a 700 hp Jumo 210G piston engine in the fuselage nose. It made the first of seven flights using this engine alone on 4 April 1941, and although underpowered, the airframe proved excellent.

Some eight months later the Me 262 V1 was fitted with a pair of early BMW 003 engines, the Jumo 210 in the nose being retained. It was doubtful whether the 990 lbs thrust produced by these units would have been sufficient to power the aircraft satisfactorily, but all speculation ended when, on 25 November 1941, the turbine blades of both engines failed at take-off revolutions.

During the summer of 1942 the third aircraft, the Me 262 V3, was fitted with a pair of 1,850 lbs thrust Jumo 004A–0 engines, and took-off successfully from Leipheim on 18 July 1942 piloted by Fritz Wendel. This was in fact the fourth pure-jet aircraft type to fly, the others being the He 178 on 24 August 1939, the He 280 on 2 April 1941 and the Gloster-Whittle E.28/29 on 15 May 1941. The Me 262 V3 was severely damaged on take-off on 11 August 1942 at Rechlin with *Dipl Ing* Beauvais at the controls.

On 2 October 1942 the Me 262 V2 replaced the third aircraft, also being powered by two Jumo 004A–0 units. The success of this aircraft prompted the RLM to increase their order for the design from 15 to 45 aircraft. In March 1943 the rebuilt Me 262 V1 joined the test programme, this being powered by two 1,850 lbs thrust Jumo 004B–0 engines. The Me 262 V4 was delivered to Lechfeld in April 1943 and on 22 May, the *General der Jagdflieger*, Adolf Galland, personally tested the machine. The fighter leader was so impressed with the new aircraft, that he called for its immediate mass production. His plea was later echoed by Göring after

The Me 262 V3 was the first aircraft to fly powered solely by turbojet engines, this event taking place on 18 July 1942.

the latter had witnessed a demonstration by the Me 262 V4 on 23 July 1943, but Hitler would only say 'Nothing will be done with the new fighter until *I* have decided on its merits'.

Following early trials, it became obvious that the tailwheel undercarriage employed by the first four prototypes afforded the pilot a far from satisfactory forward view. Therefore the fifth aircraft, the Me 262 V5, was fitted with a fixed nosewheel undercarriage and flew successfully on 23 July 1943. It also introduced mesh-like stoneguards over the intakes of the jets and provision for two 1,100 lbs thrust Rheinmetall-Borsig rocket assisted take-off units beneath the fuselage.

The Me 262 V6 (w/nr 130001) was flown for the first time in October 1943. In addition to having a fully retractable nosewheel undercarriage, the sixth aircraft was the first to be powered by two 1,980 lbs thrust Jumo 004B–1 engines. On 26 November, the Me 262 V6 was demonstrated before Hitler at Insterburg by Gerhard Lindner, Messerschmitt's new chief test pilot. Much to everyone's surprise, the *Führer* immediately saw a different use for the revolutionary fighter: 'At last there is our Blitz Bomber! Of course none of you thought of that!'.

Despite Hitler's comments, work continued on the Me 262 fighter, the V7 appearing in November 1943. This was fitted with a rubber-sealed pressurised cabin and was followed a month later by the first aircraft to carry armament, the Me 262 V8. Four 30 mm MK 108 cannon were mounted in the nose of the aircraft, which, prone to jamming as they were, packed an extremely effective punch. The Me 262 V9 (w/nr 130004) which flew in January 1944 tested various types of radio and IFF equipment. The ninth aircraft was followed, in April 1944, by a batch of 13 Me 262A–0 pre-production aircraft.

During a production conference at which Göring, Milch and Saur were present, Hitler demanded to know how many Me 262s had been completed as bombers. When told that no such variant had been built, the *Führer* flew into a rage, shouting that 'Not one of my orders has been obeyed'. Consequently, the Me 262 V10, which flew on 1 May 1944, was fitted with a pair of bomb racks.

From the early summer of 1944, the Me 262A–1a production model began to leave the assembly lines. Named *Schwalbe* (Swallow) the variant was essentially similar to the A–0s, carrying an armament of four MK 108 cannon in the nose. The Me 262A–1a/U1 was fitted with two 30 mm MK 103, two MK 108 and two 20 mm MG 151/20 cannon, and the A–1a/U2 had additional radio equipment. The Me 262A–1a/U3 was a photo-reconnaissance variant with provision for two Rb 50/30 cameras and the A–1b could carry 24 R4M rockets underwing.

During the early summer of 1944, the first Me 262A–0s were delivered to

Right: *The sixth Me 262 was fitted with a fully retractable nosewheel undercarriage. This feature was retained by all subsequent aircraft.*

Erprobungstelle Rechlin for experimental testing. Late in June 1944, the first operational jet fighter unit was established at Lechfeld near Augsburg under the designation *Erprobungskommando* 262 (Experimental unit 262). The unit was led for a short time by *Hptm* Werner Thierfelder, but he crashed to his death on 18 July after his Me 262 had burst into flames.

On 25 July 1944 a Me 262 from EK 262 recorded the world's first interception of an enemy aircraft by a jet fighter. A photo-reconnaissance Mosquito from No. 544 Squadron RAF was flying over the Munich area when the observer, F/O Lobban, spotted an enemy aircraft in the distance. The pilot F/Lt Wall, quickly accelerated the machine, but was surprised to see that the enemy was still closing rapidly. After evading five firing passes from the Me 262, Wall managed to dive into a cloud bank, eventually crash landing the Mosquito back at Fermo near Venice.

Two squadrons from III./ZG 26, which had been operating the Bf 110 and Me 410 in the defence of the *Reich*, were transferred to Lechfeld during the summer of 1944. Late in July, *Hptm* Geyer, later of *Jagdgruppe* 10, arrived to take command of these two squadrons (8. and 9./ZG 26) plus EK 262. The experienced jet pilots of EK 262 undertook the training of the other two squadrons, the latter exchanging their Me 410s for Me 262s during September and October 1944.

On 25 September the famous Austrian fighter pilot, *Maj* Walter Nowotny, relinquished command of I./JG 54 to take over a new operational unit at Lechfeld. Known as *Kommando* Nowotny, the unit became operational on 3 October 1944, with one squadron at Achmer and one at Hesepe, both near Osnabruck; a total of about 30 Me 262s. Fighter protection of the airfields was assigned to the new Fw

The first Sturmvogel *bomber prototype was the Me 262 V10, seen here towing a SD 500 bomb fitted with wings and a wheeled dolly.*

190D–9 fighters of III./JG 54 under *Hptm* Robert Weiss.

Kommando Nowotny was entrusted with the task of evolving combat tactics for use with the revolutionary fighter. Although much faster than anything that opposed it, the Me 262 was prone to engine failures, and considerable numbers were lost including eight in early November 1944. The final blow came on 8 November when Nowotny himself was killed. He had been in combat with a number of USAAF Mustang fighters when his aircraft (w/nr 110400) was seen to dive through some cloud straight into the ground. The cause of his death is still somewhat of a mystery.

Following the death of Nowotny, the *Kommando* was withdrawn to Lechfeld for refresher training and eventual incorporation into a jet fighter wing. It was at this time that *Hptm* Geyer was charged with the formation of such a wing, to be known as *Jagdgeschwader* 7 *Hindenburg*. His own unit, III./ZG 26, was combined with the remnants of *Kommando* Nowotny and eventually emerged, during the middle of December 1944, as III./JG 7.

The first commander of JG 7 was *Oberst* Johannes Steinhoff who was appointed in December 1944, leadership of III./JG 7, which became operational at Achmer, being assigned to *Maj* Erich Hohagen. Previously, in November 1944, *Maj* Theodor Weissenberger had arrived to begin the formation of I./JG 7 at Brandenburg and in February 1945, *Maj* Erich Rudorffer began the establishment of II./JG 7 at Neumünster. In January 1945, Steinhoff left JG 7, his place being taken by Weissenberger. *Maj* Heinrich Ehrler took over I./JG 7 and *Maj* Rudolf Sinner was transferred to III.*Gruppe*, now fully operational at Parchim.

The testing of the Me 262 V10 led to the production of a bomber variant, designated Me 262A–2a *Sturmvogel* (Stormbird). This aircraft could carry either one 2,200 lb SC 1000 bomb, two 1,100 lb SC 500 or two 550 lb SC 250 weapons. The Me 262A–2a/U1 had two of the MK 108 cannon deleted to make way for a TSA bombsight and the A–2a/U2 featured a semi-glazed nose to accommodate a prone bomb aimer with a *Lotfe* 7H sight. Only one Me 262A–2a/U2 was completed.

The first Me 262 bombers were delivered in June 1944 to *Kommando* Schenk at Rheine. Extensive tests were carried out by the unit, led by *Maj* Wolfgang Schenk, but only proved that the Me 262 was not really designed as a bomber! In horizontal flight, bombs had often fallen as much as two kilometres (1·2 miles) from the target. It was found that accuracy could be enhanced by releasing the bombs in a shallow dive, but for this the airframe had to be specially strengthened.

In August 1944, 3.*Staffel* of I./KG 51 under *Maj* Kurt Unrau became operational with the *Sturmvogel* at Juvincourt near Rheims. Nine aircraft were originally assigned to the unit, but by the time it arrived in France, only five remained. Late in October 1944, I./KG 51 *Edelweiss* was joined by a second group with 25 Me 262s. On 1 November 1944, Schenk, who was promoted to the rank of *Oberstleutnant*, was put in charge of KG 51, the unit occasionally operating with the less numerous Ar 234 jet bombers of KG 76.

Several other variants of the Me 262 were proposed. These included the Me 262A–3a ground-attack aircraft which featured additional armour protection and fuel tankage, the projected A–4a unarmed reconnaissance aircraft and the Me 262A–5a. This was also a reconnaissance aircraft with provision for two Rb 50/30 cameras, two MK 108 cannon and two 66 Imp gallon drop tanks.

Development work continued on the turbojets which were still far from perfect. The two main difficulties were the tendency for the units to suffer turbine failures on take-off or landing, making the aircraft almost uncontrollable. Altitude performance was also limited by an inclination for the compressor to stall and the engines to flame-out. Normal operational ceiling was limited to about 26,000 ft, although with careful handling the Me 262 could attain 35,000 ft.

One of the most novel ways of preventing turbine failure was the introduction of hollow air-cooled blades in the Jumo 004B–4 engine. This unit, which was being introduced at the end of the war, was probably used to power the extensively modified Me 262 V12 (w/nr 130007) which achieved the speed of 624 mph on 6 July 1944. Engine trouble was held responsible for 33 per cent of Me 262 crashes, another 34 per cent being attributed to the failure of the undercarriage. The Me 262 was only the second German operational aircraft to employ a nosewheel undercarriage and this, coupled with its high landing speed, rendered it particularly vulnerable to burst tyres and other more serious failures.

Although the Me 262 was a delightful machine to fly, it possessed considerably different handling characteristics from a conventional piston-engined fighter. Therefore, towards the end of 1944, the construction of a two-seat variant was proposed; the Me 262B–1a. The second seat was positioned behind the first, the whole being enclosed by an elongated transparent canopy. The installation of the second seat displaced some internal fuel tankage, this being replaced by two 66 Imp gallon drop tanks.

Following trials by a Me 262A–1a (w/nr 130056) with FuG 220 *Lichtenstein* SN–2 radar equipment, the pilot *Oberst* Hajo Herrmann pronounced the machine excellent for night fighting. Therefore, early in 1945, a radar equipped conversion of the Me 262B was produced, designated B–1a/U1. This was followed by the Me 262B–2a which featured a 'stretched' fuselage, but the aerial array carried by both machines reduced maximum speed by some 37 mph. It was hoped that the second Me 262B–2a (which was never completed) with centimetric radar in a bulged nose would considerably reduce this disadvantage.

During November a training unit was set up at Lechfeld using the personnel

Right: *A Me 262A–1a (E2+00) which was tested by Erprobungstelle Rechlin's department E2.*

The first fighter unit to operate the Me 262A–1a was Kommando *Nowotny. Here a pilot from the unit prepares for a sortie.*

from EK 262 and designated III./EJG 2. This unit operated a number of Me 262B–1as and was to be responsible for the training of ex-bomber pilots from KG 6, KG 27, KG 30, KG 54 and KG 55. In the event only I./KG(J) 54 became operational at Giebelstadt near Würzburg, the other formations being disbanded or equipped with the Bf 109. Two Me 262B–1as were also delivered to the Ar 234 training unit, III./EKG 1 at Alt Lönnewitz.

Other units equipped with the Me 262 included 12.(Erg)/KG 51, which was a training and replacement squadron for the rest of KG 51, 1./*Versuchsverband* Ob.d.L. and *Sonderkommando* Brauegg. The latter was a photographic-reconnaissance unit equipped with the Me 262A–1a/U3 which operated very successfully during the late months of 1944, providing the German High Command with valuable information on troop movements. In the spring of 1945 the unit was redesignated as the *Gruppenstab* and 2.*Staffel* of NAGr 6.

Maj Gerhard Stamp, who had previously commanded I./JG 300, began the establishment of a special Me 262 bomber destroyer unit in November 1944. During the following month, *Oblt* Kurt Welter formed a Me 262 night fighter unit for the defence of Berlin. Initially designated *Kommando* Welter, the unit is reported to have destroyed 20 Allied aircraft. In April 1945 it was redesignated 10.*Staffel*/NJG 11.

Some of the most spectacular developments of the basic Me 262 were those in which rocket engines were used to boost the fighter's climbing performance. The first variant was the Me 262C–1a which was fitted with a 3,750 lbs thrust Walter 509 rocket unit in the tail. The first aircraft

The Me 262A–1b carried a rack for twelve 55 mm R4M rockets under each wing.

(w/nr 130186) made its initial flight on 27 February 1945 and eventually attained an altitude of 38,400 ft in 4·5 minutes. The second variant was designated Me 262C–2b and differed in being powered by two BMW 003R engines. These units comprised a standard BMW 003 turbojet coupled with a BMW 718 rocket engine, the total producing almost 4,000 lbs thrust for 2 minutes. Only one flight was made with the Me 262C–2b (w/nr 170078) piloted by Karl Baur, it being estimated that the aircraft could attain an altitude of 42,500 ft in 2 minutes 20 seconds. The final rocket-boosted variant was the Me 262C–3a which was to carry a jettisonable Walter 509 engine beneath the fuselage.

Some of the most advanced variants of the Me 262 were the *Aufklärer* (reconnaissance) and *Schnellbomber* (fast-bomber) projects. The first variant was similar to the basic Me 262A–1a, but the second (designated *Aufklärer* Ia or *Schnellbomber* Ia) had the cockpit moved forward to the extreme nose. In contrast to this very attractive design, the third project was extremely ugly, featuring a bulged fuselage belly to accommodate bombs or fuel internally. The final, and most fascinating project was for the 3 *Sitziger Nachtjäger*

After the war a number of Me 262s were gathered together for evaluation by the Allies. Here an Me 262A–2a bomber from KG 51 is standing in front of a JG 7 Me 262A–1a.

(3-seat night fighter). This was developed from the Me 262B–2a but featured a fuselage lengthened by 6½ feet to accommodate a third crew member, wings swept back by 35 degrees and two 2,860 lbs thrust HeS 011 turbojets buried in the wing roots.

The last and perhaps most famous Me 262 unit was *Jagdverband* 44 which was formed on 10 February 1945 from a nucleus provided by 16./JG 54. Commanded by *Gen* Adolf Galland, the unit included some of the most famous pilots of the *Luftwaffe*. Aces such as *Oberst* Johannes Steinhoff, *Oberst* Günther Lutzow, *Oberstlt* Heinz Bär, *Maj* Gerhard Barkhorn, *Maj* Walter Krupinski, *Maj* Karl-Heinz Schnell and *Maj* Erich Hohagen all served with the unit.

On 18 March 1945, the USAAF launched the heaviest attack on Berlin of the war. It was intercepted by Me 262s from I. and II./JG 7 and the destruction of 19 heavy bombers was claimed. After a period of working-up at Brandenburg, JV 44 became operational at München-Riem with a nominal strength of 25 aircraft. On 4 April, 49 fighters from JG 7 intercepted an American bomber formation over Nordhausen. The commander of III.*Gruppe*, *Maj* Sinner, was shot down by Mustangs on take-off from Parchim and wounded, his place being taken by *Maj* Späte from JG 400.

Both JG 7 and JV 44 were equipped with 55 mm R4M rocket firing Me 262s. The use of these missiles had been pioneered by *Maj* Sinner, *Maj* Georg Cristl of JGr 10 and *E-Stelle* Rechlin and proved very effective. When attacking, the Me 262 would approach the bomber stream in a 35 degree dive, then pull up into a sharp climb to reduce speed before releasing its R4M missiles and opening fire with its cannon.

Prior to 7 April 1945, all jet fighter operations had been directed against enemy bombers, but on that day, under the code name of 'Wehrwolf', JG 7 attacked the fighter escort. It was widely held that the tactic should have been adopted earlier, leaving the bombers to the mercy of the heavily armed piston-engined fighters. On 10 April, jet fighter bases were heavily bombed and several

Me 262s destroyed. This forced many units to more distant airfields in Austria and Czechoslovakia, and in one case to utilise an autobahn in which the central reservation had been concreted over.

From this date until the final collapse, jet operations were carried out amidst growing difficulties. On 18 April, Steinhoff was severely burnt when his Me 262 crashed on take-off. Six days later Günther Lützow was killed, and on 26 April 1945, *Gen* Galland was injured. On the same day, 95 Me 262s were flown into München-Riem from defunct jet units for use by JV 44, but by then it was too late. *Oberstlt* Heinz Bär, who had previously brought his He 162 unit, EK 162, to join JV 44, took over command and supervised its transfer to Salzburg-Maxglam. When the Americans entered the beautiful Austrian city on 3 May 1945 all that remained of JV 44 were a few burning Me 262s.

Specification (Me 262A–1a)

SPAN	12·50 m	(41 ft 0½ ins)
LENGTH	10·61 m	(34 ft 9½ ins)
HEIGHT	3·83 m	(12 ft 6¾ ins)
EMPTY WEIGHT	4,000 kg	(8,820 lbs)
LOADED WEIGHT	6,775 kg	(14,938 lbs)
MAXIMUM SPEED	868 km/h	(536 mph)
NORMAL RANGE	845 km	(525 miles)

Above: *The two-seat Me 262B–1a trainer which survived until recently in Czechoslovakia.*

Right: *Several Me 262A–1a fighters were built after the war in Czechoslovakia, intended for use by that air force.*

Me 263

DESPITE possessing a remarkable performance, the Me 163 *Komet* had a number of serious drawbacks amongst which were its exceedingly short range and difficulties experienced when attempting a landing with its skid type undercarriage. Therefore, early in 1944, Messerschmitt began work on an improved variant initially designated Me 163D. The aircraft introduced an entirely new fuselage with increased fuel capacity and a conventional retractable nosewheel undercarriage.

At an early stage in its development, the Me 163D project was taken over by Junkers as the Ju 248, and completed under the direction of Prof Heinrich Hertel. The prototype was completed at Dessau in August 1944, and was again redesignated, becoming the Me 263 V1. The aircraft carried a total of 352 Imp gallons of T-*Stoff* and 185 gallons of C-*Stoff*, most of the latter in wing tanks. Apart from the innovations already noted, the Me 263 was fitted with a pressurised cockpit, air for which was provided by a small compressor driven by a three-bladed airscrew in the nose. The compressor maintained an altitude pressure of 26,250 ft up to an actual height of almost 50,000 ft.

The wing of the Me 263 V1 was substantially similar to that of the Me 163B, but the fixed leading edge slots of the latter were replaced by automatic devices in the former. Despite the increase in fuselage length, no tailplane was fitted and the fin and rudder were essentially similar to those of the previous design. Like the improved Me 163C model, the Me 263 was powered by the 4,430 lbs thrust Walter 109–509C rocket engine and carried an armament of two 30 mm MK 108 cannon in the wing roots. Total ammunition capacity was increased to 150 rounds and the armour protection provided for the pilot was revised.

Initial unpowered flight trials with the Me 263 V1 were carried out by Junkers, the machine's undercarriage being locked down. Late in 1944, the aircraft was handed back to Messerschmitt for powered flight trials. Although successful, the Me 263 proved somewhat of a disappointment. At speeds in excess of Mach 0·8, the aircraft's centre of pressure varied so rapidly that the throttle had to be eased back at that point. The Me 163's limiting Mach number was 0·82. Consideration was given to the design of a tailplane, but development of the basic variant continued under the designation Me 263A–1. In the event, no production aircraft were delivered before the end of the Second World War.

Specification (Me 263A–1)

SPAN	9·50 m	(31 ft 2 ins)
LENGTH	7·88 m	(25 ft 10½ ins)
EMPTY WEIGHT	2,000 kg	(4,410 lbs)
LOADED WEIGHT	5,300 kg	(11,660 lbs)
MAXIMUM SPEED	950 km/h	(590 mph)
ENDURANCE	15 mins	(15 mins)

Left: *Only one prototype of the Me 263 rocket propelled fighter was completed before the end of the war.*

Me 264

'NOW THAT we know the right line is to produce revolutionary inventions and leave botches and bungles alone, I am all in favour of going all out with the Me 264. This Me 264, with its combination of piston engines and turbojets offers us prospects of revolutionary importance. Its load and range make it seem adapted to all purposes, even in relatively small numbers.'

The words of Karl-Otto Saur, armament supervisor and head of the *Jagerstab* spoken in June 1944, clearly illustrate that the Me 264 was being considered for production even at that stage in the war. The Me 264 had been conceived by Messerschmitt in 1940 as a private venture for an ultra long-range heavy bomber able to strike at cities on the American east coast. Eventually the RLM foresaw the propaganda value of such a project and asked Messerschmitt, Focke-Wulf and Junkers to proceed with detailed design under a low development priority. Besides the Messerschmitt design, Focke-Wulf produced the Fw 300 project based on the *Condor* and Junkers a development of the Ju 290.

Eventually, Messerschmitt obtained permission from the RLM to go ahead with the construction of three prototypes of the bomber. The Me 264 was an extremely clean shoulder wing monoplane with twin tails and powered by four 1,340 Junkers Jumo 211J engines in neat annular cowlings. The Me 264 V1 was completed late in 1942, being flown for the first time in December. The aircraft carried a total of 4,330 Imp gallons of fuel and a bomb load of almost 4,000 lbs was proposed. No defensive armament was carried and the design was unusual at this time in featuring a nose-wheel undercarriage.

The four-engined Me 264 was known unofficially as the Amerika-Bomber *being designed to attack New York from European bases.*

Following the entry of the United States into the war, the RLM demanded that the ultra long-range bomber projects should be fitted with six engines and defensive armament. This led the three manufacturers to produce the Focke-Wulf Ta 400, the Junkers Ju 390 and the Messerschmitt Me 264B. Of the three designs, the RLM favoured the Ju 390 and asked Messerschmitt to develop the Me 264 into a long-range maritime reconnaissance bomber.

Therefore, the Me 264 V2 was powered by four 1,700 hp BMW 801D or G radial engines, extensive armour protection and provision for defensive armament and cameras. The second prototype was being readied for flight testing when it was destroyed in a bombing attack late in 1943. Work continued on the third aircraft, this like the V2, featuring an increase in wing span from 127 ft 7½ ins to 141 ft 1 in. Two dorsal gun turrets were fitted, the forward position carrying a 13 mm MG 131 machine gun, the after position a 20 mm MG 151 cannon. MG 131 weapons were also installed in the nose, ventral and beam

The Me 264 V1 shows off its attractive lines during an early flight test.

positions and provision was made for 4,400 lbs of bombs and three Rb 50/30 cameras.

The Me 264 V3 was to have carried a crew of eight and six 2,200 lb rocket assisted take-off units were proposed to assist the bomber in getting airborne in an overloaded condition. Despite Saur's optimistic words in June 1944, the Me 264 was abandoned soon afterwards, the V3 probably never being completed.

Before the aircraft was eventually abandoned, several more advanced variants were proposed. Two 1,980 lbs thrust Jumo 109–004 or two 7,700 lbs thrust BMW 109–018 turbojets were to be added to the normal piston engines to increase cruising speed, and another variant was proposed powered only by two 5,440 hp BMW 109–028 turboprop engines. The version with two boosting turbojets was to have its tailplane mounted on top of a single fin to enable it to clear the jet efflux. Another project was to have involved the use of Ritz heat exchangers to greatly increase range.

Following the plot on Hitler's life in July 1944, the Me 264 V1 had been readied at Lechfeld to fly the *Nazi* leader to Japan should the rebellious Generals gain the upper hand. It was later delivered to *Transport-Fliegerstaffel* 5, although it is very doubtful whether the aircraft was flown by the unit before being used as a test bed for the Osermaschinen steam turbine project.

This was undoubtedly the most unusual power plant proposed for the Me 264. The turbine was intended to develop 6,000 hp and drive a 17½ foot diameter airscrew, fuel being in the shape of a mixture of powered coal and petroleum. Main advantages of the steam turbine were constant power at varying altitudes, simple maintenance and immunity to low temperature troubles. Although the engine was to be installed in the Me 264 V1 the war ended before any tests could be made.

Specification (Me 264 V3)

SPAN	43·00 m	(141 ft 1 in)
LENGTH	20·90 m	(68 ft 7 ins)
EMPTY WEIGHT	23,360 kg	(51,509 lbs)
LOADED WEIGHT	45,540 kg	(100,416 lbs)
MAXIMUM SPEED (with GM–1)	565 km/h	(351 mph)
MAXIMUM RANGE	15,000 km	(9,315 miles)

Me 309

THROUGHOUT THE WAR, much energy was expended by the Messerschmitt company in attempting to produce an improved version of the famous Bf 109. One of the most important of these was the Me 309, a DB 603 powered fighter with a nosewheel undercarriage and a retractable radiator.

Prior to the completion of design work, several of the aircraft's systems were tested by converted Bf 109s. The retractable radiator was fitted to the Bf 109 V24 (w/nr 5642) for windtunnel tests at Göttingen and to the V31 (w/nr 5642) for flight trials. Two pressurised cabin designs were tested, respectively by the Bf 109 V30 (w/nr 5716) and the V30a (w/nr 5717) and a nosewheel undercarriage was fitted to a Bf 109F–1 (w/nr 5603).

Nine prototypes of the Me 309 were ordered, the first of which began taxying trials at Augsburg-Haunstetten on 27 June 1942. During these tests, severe nosewheel wobble caused a fracture in the pipes leading from the radiator to the engine. Eventually the Me 309 V1 became airborne on 18 July 1942, but this flight only lasted seven minutes because of severe overheating of the engine and lack of sufficient hydraulic power to retract the undercarriage. Subsequent flight trials resulted in a number of modifications, the aircraft resuming flight trials on 5 August 1942. Despite these modifications the Me 309 was still not particularly successful. One of the greatest disappointments was that the machine still manifested a similar take-off swing to the Bf 109. At least five different tail fin and rudder profiles were designed for the machine in an attempt to cure this, but with little success.

The Me 309 V3 joined the test programme on 28 November 1942; many minor crashes being experienced by both prototypes. Cannibalisation of the subsequent aircraft led to the number of prototypes required being reduced from nine to four. Eventually the Me 309 V2 broke its back in a severe crash, but was replaced early in 1943, by the V3.

Previously, in November 1942, the Me 309 V1 was transferred to Rechlin and test flown by *Dipl Ing* Beauvais. The test pilot was not particularly impressed by the machine and considered that it had particularly high stick forces. This, combined with the fact that the Bf 109G could easily out-turn the Me 309, led to the new fighter's abandonment in late 1943.

The Me 309 V4 was completed in July 1943, being used as an armament test aircraft. The machine was fitted with two 13 mm MG 131 machine guns in the fuselage, two similar weapons plus two 20 mm MG 151 cannon in the wing roots and one 30 mm MK 108 gun in each leading edge. The Me 309 V4 was destroyed in a bombing raid on Leipheim airfield.

Two production versions had been proposed, the Me 309A–1 fighter and the Me 309B–1 dive bomber with provision

Right: *The Bf 109 V31 (w/nr 5642) flight tested the inwards retracting undercarriage and radiator designed for the Me 309.*

for two 550 lb bombs. Apart from the standard 1,750 hp DB 603A engine versions, a development was proposed powered by the improved DB 603H unit and another with the Jumo 213. The Me 409 was a project for a slightly larger version of the Me 309, no Me 509 has yet been traced, and the Me 609 was to have comprised two Me 309 fuselages coupled together by a new centre section.

Specification (Me 309 V1)

SPAN	11·04 m	(36 ft 2¾ ins)
LENGTH	9·46 m	(31 ft 0½ ins)
EMPTY WEIGHT	3,530 kg	(7,784 lbs)
LOADED WEIGHT	4,250 kg	(9,371 lbs)
MAXIMUM SPEED	733 km/h	(455 mph)
NORMAL RANGE	1,400 km	(870 miles)

Me 321 Gigant

DURING EARLY 1940, the RLM displayed interest in the design of a large transport glider following the success attained by the DFS 230. Two companies were asked to begin work on designs: Junkers who produced the Ju 322 *Mammut* (Mammoth) and Messerschmitt who designed the smaller Me 321 *Gigant* (Giant).

The Me 321 was designed by *Dipl Ing* Degel and was basically a hugh high-wing monoplane of metal structure with fabric covering. An interesting feature was the multi-wheel undercarriage designed to assist landings in rough terrain. The first prototype, the Me 321 V1, made its initial flight in March 1941 from Leipheim towed by a Junkers Ju 90. Despite displaying an ability for easy floating landings, the Me 321 was by no means a simple aircraft to control. Extremely high stick forces without the benefit of power assisted controls gave the pilot an extremely hard job.

After a tracked vehicle had fallen through the floor of the Ju 322, the *Gigant* was chosen for production. Typical loads carried by the Me 321 included a company of infantry, 22 tons of freight or a small tracked vehicle. Loading was facilitated by large 'clam shell' doors in the nose of the aircraft or two smaller doors on either side of the fuselage. Two production variants were built, the Me 321A–1 which carried a single pilot, and the Me 321B–1 which had a crew of three and was fitted with an armament of four 7·9 mm MG 15 machine guns.

Various methods of towing the Me 321 were employed. The four-engined Junkers Ju 90 airliner was hardly powerful enough for the job, and the novel scheme of employing three Bf 110s in vee formation was tried. Known as the *Trioka-Schlepp*, this method proved reasonably satisfactory, although the ideal towing vehicle was the five-engined Heinkel He 111Z. Apart from using powered aircraft as a means of getting the Me 321 off the ground, several schemes for rocket assisted take-off were used. These included eight 1,100 lb

Left: *The Me 309 V1 was flown initially without its spinner which resulted in poor airflow to the retractable radiator.*

Facing page: *A pair of Me 321A* Gigant *glider transports at Leipheim during 1941.*

Walter 109–500, four 2,200 lb Walter 109–501 or four 3,300 lb Walter 109–502 rocket units.

The first operational unit to receive the glider was the *Grossraumlastensegler-Gruppe* Me 321 (high capacity transport group Me 321) based at Leipheim under Fritz Morzik. The unit comprised three *Staffeln* of six Me 321s each, together with a *Motorschleppstaffel* (powered towing squadron) of four groups of three Bf 110s. The unit was formed with the intention of transporting material to England following the proposed invasion, but was disbanded in December 1941. The aircraft later equipped the *Grossraumlastensegler-Kommando* 2 which was formed at the Messerschmitt plant at Obertraubling near Regensburg. This unit operated many supply missions in Russia during the 1943–44 period.

Between March 1941 and April 1942 when production of the Me 321 ceased, 175 aircraft were completed. Apart from the more conventional methods of towing and rocket assisted take-off, at least one Me 321 was fitted with a number of 660 lbs thrust Argus As 014 impulse duct engines underwing. Development of the Me 321 led to the Me 323 powered transport.

Specification (Me 321B–1)

SPAN	55·00 m	(180 ft 5½ ins)
LENGTH	28·15 m	(92 ft 4¼ ins)
HEIGHT	10·15 m	(33 ft 3½ ins)
EMPTY WEIGHT	12,400 kg	(27,342 lbs)
LOADED WEIGHT	34,400 kg	(75,852 lbs)
NORMAL TOWING SPEED	160 km/h	(99 mph)

Me 323 Gigant

FOLLOWING the completion of the first Me 321 glider, *Dipl Ing* Degel began the development of two powered variants under the tentative designations Me 321C and D. The aircraft were to be structurally strengthened and fitted with four or six captured 1,140 hp Gnôme-Rhône 14N 48/49 engines. It was also proposed that the variant be fitted with new twin fin and rudder surfaces, but in the event the original unit was retained.

During the autumn of 1941, the Me 321 C and D were redesignated as the Me 323C and D, a prototype of each being converted from a Me 321B under the respective designations Me 323 V1 and V2. The former was powered by four French engines, the latter by six. Both prototypes were tested at Lechfeld during the spring of 1942, the four-engined variant being quickly abandoned. The Me 323 V2 was followed by a batch of ten pre-production Me 323D–0s which bore the experimental designations V3 to V12. The Me 323D–0 differed from the early aircraft in having less cabin windows and the tailskid moved further aft.

The production Me 323D–1 carried an armament of five 7·9 mm MG 15 machine guns in the nose which could be supplemented by six MG 34 infantry weapons of similar calibre fired through ports in the fuselage sides. The Me 323D–2 was similar apart from some redesign of the engine

Left: *A Me 321A glider takes-off under the power of eight 1,100 lb thrust Walter rockets at Orsha in the Soviet Union.*

The Me 323 V2 was the first aircraft to be powered by six engines. It was the forerunner of the Me 323D series.

cowlings and the provision of two-blade wooden airscrews. The Me 323D–6, which appeared in December 1942, carried four MG 131 and two MG 15 machine guns and some further modification to the engine cowlings.

Previously, in February 1942, a more advanced version of the Me 323 had been proposed, powered by six 1,600 hp BMW 801 radial engines and featuring increased armament. Despite its promise, the variant was abandoned because of the shortage at that time of the new BMW engines needed for more important types such as the Fw 190A and Do 217E.

The first unit to receive the powered *Gigant* was I./KGzbV 323 (Bomber Wing on Special Operations 323) formed towards the end of October 1942. Equipped with nine aircraft, the unit became operational in the Mediterranean area, ferrying supplies to the North African front. It was later joined by II./KGzbV 323, the two units attempting, on 22 April 1943, to ferry petrol from Tunisia for use by Rommel's troops. They were intercepted over the Gulf of Tunis by a large formation of RAF fighters and 14 Me 323s crashed in flames. During the month of April KGzbV 323 was decimated and withdrawn from operations.

In May 1943, the two Me 323 groups were reconstituted as *Transportgeschwader* 5 under *Oberst* Gustav Damm. (The proposed formation of III./KGzbV 323 from KGrzbV 900 was never completed.) I./TG 5 was commanded by *Maj* Günther Mauss and II./TG 5 was led by *Oberstlt* Stephan. After its re-establishment, TG 5 was transferred to the eastern front.

Early in 1943, the Me 323 V13 and V14 were completed. Both aircraft featured improved defensive armament comprising 13 mm MG 131 machine guns, the former being powered by six Gnôme-Rhône engines, the latter by six 1,340 hp Jumo 211F units. The Me 323 V13 was developed into the similar Me 323E–1 production model. The Me 323 V15 was fitted with two HDL 151 gun turrets each of which housed a 20 mm MG 151 cannon, but the increased drag produced by these positions had an adverse effect of the aircraft's performance.

The Me 323E–2 featured low-drag EDL 151 turrets and an experimental version, the E–2/U1, carried no less than eleven MG 151 cannon and four MG 131 machine guns. Powered by six Jumo 211F engines, the E–2/U1 proved too cumbersome for operational service. The Me 323 V16 was prototype for the proposed Me 323F–1 powered by six 1,350 hp Jumo 211R engines, but the variant failed to see service.

By the beginning of 1944, the Messerschmitt company was concentrating all its production facilities on the building of the Bf 109 and introducing the Me 262 jet fighter into service. Therefore it was decided to transfer production of the Me 323 to the Zeppelin works at Friedrichshafen. Work then began on a final variant to be built by the Zeppelin company, the Me 323G. This was to be powered by six 1,320 hp Gnôme-Rhône 14R engines. A prototype, the Me 323 V17 was under construction at Leipheim when the *Gigant* was finally abandoned.

Further proposed variants were the ZMe 323H which, based on the Me 323E, was to carry a much increased load with a consequent reduction in speed to 150 mph. The ZMe 423 was a project based on the ZMe 323F–1 and the ZSO 523 was a further developed version of the ZMe 323G with a butterfly tail to be produced jointly by the Zeppelin and Sud Est company of France.

By this time however, the Me 323 was becoming obsolete, being gradually replaced by such aircraft as the Ar 232, the Ju 352 and the Ju 390. Production of the *Gigant* ended in March 1944; 27 aircraft

Right: *The Me 323 V3 was the first of a batch of ten Me 323D–0 pre-production aircraft.*

being built in 1942, 140 in 1943 and 34 in 1944, a total of 201 machines.

Early in 1944 the Me 323s of *Transportgeschwader* 5 were placed under the control of the general transport force on the Russian front so that they could provide air support where it was most needed. By March 1944, the Arado Ar 232 and Italian Savio-Marchetti SM 82 had begun to replace the Me 323 in service with TG 5 and following the invasion of Normandy in June 1944 the complete *Geschwader* was disbanded.

Specification (Me 323D–6)

SPAN	55·00 m	(180 ft 5½ ins)
LENGTH	28·15 m	(92 ft 4¼ ins)
HEIGHT	9·60 m	(31 ft 6 ins)
EMPTY WEIGHT	27,330 kg	(60,260 lbs)
LOADED WEIGHT	43,000 kg	(94,815 lbs)
MAXIMUM SPEED	285 km/h	(177 mph)
NORMAL RANGE	1,100 km	(683 miles)

The Me 323D–1 was the first major production model. It was powered by six 1,140 hp Gnome-Rhone 14N 48/49 radial engines.

Above: *The badge below the cockpit of this Me 323E–1 indicates that this aircraft was operated by II.*Gruppe/*TG 5.*

Right: *The Me 323E–2/U1 was powered by six 1,200 hp Jumo 211 engines and had an HDL 131 gun turret in the extreme nose.*

Me 328

IN ADDITION to building the world's first operational jet and rocket-propelled aircraft, the Messerschmitt company also produced the only fighter designed to be powered by pulsejet engines, the Me 328.

Much of the early work on pulsejet engines was carried out by Paul Schmidt, although the first official German contract for such a unit was placed with the Argus company in 1939. Early in 1940 the design team, under Dr Ing Fritz Gosslau, heard of Schmidt's work and were allowed to examine one of his early engines. The result of this collaboration was the testing beneath a Go 145 (D–IIWS) of a 265 lbs thrust unit on 30 April 1941. These trials prompted Gosslau to begin work on a more powerful unit which eventually emerged as the Argus As 014.

In July 1941, Messerschmitt began feasibility studies on a small expendable pulsejet powered aircraft under the designation P.1079. In March 1942 these proposals crystallised as the P.1079/17 or Me 328; an escort fighter or fighter-bomber to be towed behind such aircraft as the He 177 or Me 264. Six variants were proposed; the Me 328A–1 with an armament of two 20 mm MG 151 cannon, the A–2 with increased wing area and an extra pair of 30 mm MK 103 cannons and the A–3 which could be re-fuelled in flight. The Me 328B–1, B–2 and B–3 were similar but designed for the fast bomber rôle.

It was not until March 1943 that serious development was undertaken when it was realised that the Me 328 would make a useful low-level bomber should the Allies attempt an invasion of France. Develop-

The all wooden Me 328 V1 took part in gliding trials at Hörsching near Linz in Austria.

The Jacob Schweyer Segelflugzeugbau joined with the Messerschmitt company and DFS in the development of the Me 328.

ment was handed over to the Jacob Schweyer Segelflugzeugbau, the company working in co-operation with both DFS and Messerschmitt.

Messerschmitt was responsible for the aerodynamic research on the machine, commencing a series of trials in their Augsburg wing tunnel to determine the best position for the pulsejet engines. At first it was proposed to mount the engines on either side of the fuselage, exhausting beneath the tailplane. Although theoretically the best position, it was found that the power plants would set up severe vibrations around the rear of the all wooden airframe. Eventually it was decided to position the engines beneath the wings on flexible mountings.

The first three prototypes were built by the DFS research institute, all featuring metal fuselages, wooden wings and tailplanes taken from standard Bf 109 fighters. The Me 328 V1 was mounted on the back of a Do 217E and commenced tests from Hörsching near Linz during the autumn of 1943. After a series of tethered flights, the machine was released from various altitudes to enable it to undertake gliding

Above: *The Me 328B was powered by two 660 lb thrust Argus As 014 impulse duct engines and landed with the aid of a retractable skid.*

Right: *An early Me 328 prototype mounted above a Do 217 bomber for engine trials.*

trials, the Me 328 V2 and V3 being similarly employed.

At an early stage in the development programme it had been decided to concentrate work on the Me 328B. Originally designed as a conventional fast bomber, the Me 328 was later developed as a piloted missile for pinpoint attacks on heavily defended targets. A further seven prototypes (Me 328 V4 to V10) were completed by the Jacob Schweyer company, these being built entirely of wood and powered by a pair of 660 lbs thrust Argus As 014 pulsejet engines.

Trials indicated that flight characteristics were far from perfect, and the vibrations set up by the crude form of propulsion did little to help matters. On several occasions the pilot was forced to jettison the engines and two aircraft disintegrated during flight trials. Take-off was made with the assistance of a Madelung KL 12 catapult or a Rheinmetall-Borsig rocket-propelled rail trolley, the aircraft landing on a retractable skid.

Construction of a pre-production batch of Me 328B–0 bombers was commenced by Jacob Schweyer during 1944. The aircraft could carry a 2,205 lb or 1,102 lb bomb externally, mounted below the retractable skid. Some armour protection was provided for the pilot, and it was possible to jettison the complete rear fuselage in order to facilitate easy escape from the aircraft in an emergency. The fuselage contained four 110 gallon fuel tanks, and the wings were fitted with balloon cable cutters along the complete leading edge to assist in the penetration of heavily defended targets.

The final version of the aircraft to be projected was the Me 328C in which the pulsejets were to be replaced by a 1,980 lbs thrust Jumo 004 turbojet mounted in the rear fuselage. Although this design might have proved effective, it largely nullified the Me 328's main advantage; cheapness and quickness of construction. It had been estimated that four Me 328s could have been produced in the same time that it took to build a Fw 190 piston-engined fighter.

In the event, the Me 328B never saw operational service although it had been planned at one time to issue the machine to 5./KG 200 under *Hptm* Heinrich Lange. Often known by the code-name '*Leonidas Staffel*', the unit was to use the Me 328 as a manned missile, the pilot baling out just before the aircraft struck the target. The many troubles besetting the Me 328 led to a switch to piloted versions of the V–1 flying bomb and a final abandonment of the Messerschmitt design.

Specification (Me 328B–1)

SPAN	6·40 m	(21 ft 0 ins)
LENGTH	6·83 m	(22 ft 4¾ ins)
HEIGHT	2·50 m	(8 ft 2½ ins)
LOADED WEIGHT	2,700 kg	(5,940 lbs)
MAXIMUM SPEED	680 km/h	(423 mph)
RADIUS OF ACTION	630 km	(392 miles)

Me 410 Hornisse

As has been described in the Me 210 history, the failure of the original design led to a series of both major structural and detailed systems modifications being undertaken. Despite all this work, the Me 210 never really emerged as a successful aircraft and in an attempt to salvage something from its failure, Messerschmitt developed the re-engined Me 410 *Hornisse* (Hornet) in 1942.

Original proposals had centred around the high-altitude Me 310 project which was to introduce a high aspect ratio wing with a span of 58 ft 8¾ ins. The project was to be powered by two 1,750 hp Daimler-Benz 603A engines driving four-blade airscrews. The new wing was fitted to a Me 210 (w/nr 179) which made its first flight on 11 September 1943. Two production versions were proposed, the Me 310A–1 fast bomber and the A–2 destroyer, but the Me 310 was abandoned in favour of the less radically modified Me 410.

This was basically a development of the latest Me 210 production aircraft powered by two 1,750 hp Daimler-Benz DB 603A engines. Externally, the *Hornisse* could only be distinguished from its predecessor by an eight-inch increase in the length of the engine cowlings. Following the modification of six Me 210As the first true prototype, the Me 410 V1, flew towards the

Above: *The Me 210 V13 which was fitted with a pressurized cockpit and four-bladed propellers in connection with the Me 310 programme.*

Right: *The first true prototype for the* Hornisse, *the Me 410 V1, flew towards the end of 1942.*

end of 1942, followed by the generally similar V2 and V3.

Three basic production models were proposed: the Me 410A–1 light bomber, the A–2 destroyer and the A–3 reconnaissance aircraft. The first two variants carried similar armament installations to those of the Me 210A–2 and A–1 respectively, the Me 410A–3 having the forward-firing MG 17 machine guns deleted.

A large variety of minor modifications of the basic designs were also produced. The first of these was the Me 410A–1/U1 photographic reconnaissance aircraft with the MG 151/20 cannons deleted and provision for an Rb 20/30, Rb 50/30 or Rb 75/30 camera. The Me 410A–1/U2 carried two extra MG 151/20 cannon in the bomb bay and the A–1/U4 carried a 50 mm BK 5 gun under the fuselage. This huge weapon carried 21 rounds and was used operationally by II./ZG 26, although its bulk severely restricted manoeuvreability. Many Me 410A–2s were fitted with 30 mm MK 108 cannon in place of the MG 17 machine guns, the A–2/U2 variant being a night fighter with *Lichtenstein* radar equipment. The A–2/U1 and A–2/U4 had similar modifications to those of the A–1/U1 and A–1/U4.

The Me 410A was followed by the B-series which differed in being powered by two 1,900 hp DB 603G engines. The Me 410B–1 was a light bomber and the B–2 was a destroyer. The Me 410B–1/U2 and U4 had similar modifications to the A–1/U2 and U4, but the Me 410B–2/U4 carried two 30 mm MK 103 cannon in addition to the 50 mm BK 5 gun.

In addition to these modifications, a series of Rüstsatz conversion packs were produced for use with the Me 410B series. The Rüstsatz R1 was an extra pair of 20 mm MG 151 cannon in a WB 151 installation, the R2 introduced two 30 mm

The Me 410 V3 (w/nr 055) was generally similar to the first two prototypes.

A standard production Me 410A (w/nr 10047) coded 5T+DN of an unidentified Luftwaffe *unit. This aircraft may have been rebuilt from a Me 210.*

MK 108 guns, and the R3 comprised two MK 103 weapons of similar calibre. The MK 108 was a light weapon built largely of welded steel plate with a rate of fire of 660 rounds per minute. The MK 103 was more than twice as heavy, had considerably increased range and accuracy, but suffered from the slower rate of fire of 420 rounds per minute. The R4 conversion pack comprised two MG 151 cannon in a jettisonable WT 151 installation.

Apart from these more conventional weapons, several Me 410s were fitted with WGr 21 rocket tubes, the 210 mm weapon usually being mounted one under each wing. At least one aircraft was fitted with four WGr 21 tubes in a rotating mounting installed in the bomb bay. A project also existed for fitting the Me 410 with the huge 75 mm BK 7·5 gun although it is not known whether any such aircraft was so modified. The BK 7·5 carried 12 rounds and was fitted to the He 177A–3/R5 and Hs 129B–3, and in its earlier form, to the Ju 88P–1.

Further B-series aircraft included the Me 410B–3 reconnaissance aircraft similar to the A–3 apart from its engines, and the B–5 experimental torpedo-bomber. Some of these machines had the gun barbettes removed to make way for extra fuel, and several advanced weapons were tested by the aircraft. The Me 410B–6, which was used by I./ZG 1 at Lorient and Vannes, was an anti-shipping aircraft equipped with FuG 200 *Hohentwiel* search radar. The Me 410B–7 and B–8, day and night reconnaissance aircraft respectively, were not built.

The Me 410C was a projected high-altitude variant powered by DB 603JZ engines in annular cowlings. The machine was fitted with a similar high aspect ratio wing to that of the Me 310, but was not completed. The generally similar Me 410D with wooden outer wing panels was abandoned in favour of the Me 410H high-altitude fighter. This was to have an even larger wing with a span of 75 ft 5 ins. Powered by DB 603G engines and having some modification to the cockpit glazing, the Me 410H was not built.

In May 1943 the *Hornisse* began to replace the Me 210 in service with 2.(F)/122 at Trapani and III./ZG 1 at Gerbini. About the same time, 5./KG 2 took several Me 410s on hand, but passed these machines to V./KG 2 after the latter was formed from II./KG 40 and KG 101. The fifth *Gruppe* of KG 2 became operational during the summer of 1943, its first loss coming on 13 July when 16.*Staffel* lost w/nr 238 to a Mosquito of No. 85 Squadron, RAF.

One of the most interesting units formed within the *Luftwaffe* was *Erprobungs-kommando* 25, which was charged with the task of introducing new anti-bomber weapons to operational service. Established in April 1944, the unit was commanded by *Hptm* Eduard Tratt, an experienced destroyer pilot who had previously commanded 1./ZG 1. It comprised three squadrons: a *Jagdstaffel*, a *Kampfstaffel* and a *Zerstörerstaffel*. The latter, equipped with a mixture of Bf 110s and Me 410s was by far the most important, being responsible for the development of subsequent destroyer unit tactics. The *Staffel* tested the WGr 28 rocket tubes, the 30 mm MK 101 cannon, and the 37 mm BK 3·7 and 50 mm BK 5 guns.

In September 1943, Tratt left *Erprobungskommando* 25 to take over command of II./ZG 26 *Horst Wessel* which had just re-equipped with the Me 410A. Based at Oberpfaffenhofen, the unit suffered much from the attentions of USAAF long-range fighter patrols, but eventually became operational in the defence of Germany, operating from Königsberg/Oder. The unit was normally escorted by the single-engined fighters of JG 300, its task being to penetrate the Allied fighter screen and destroy the raiding bombers with its heavy armament.

On 22 February 1944, Tratt, now a *Maj* was killed near Nordhausen/Harz when flying Me 410 (w/nr 420410). He was attempting a foolhardy single-handed attack on a USAAF bomber formation. Tratt was the leading destroyer pilot having flown 350 missions, destroyed 38 Allied aircraft and 24 tanks, and awarded the Knight's Cross with Oak Leaves. The only eye-witness to Tratt's death was *Oblt* Prokopp who was killed a short while later when his Me 410 rammed a P–47 Thunderbolt fighter.

Above: *The Me 410A–2/U4 carried a 50 mm BK 5 cannon under the fuselage. Although severely limiting manoeuvreability, the BK 5 equipped* Hornisse *was used operationally by ZG 26.*

Right: *For exactly two years the Me 410A–3 was used by 2.(F)/122 in Italy, the type serving longer with this unit than any other.*

In April 1944, II./ZG 1 under *Hptm* Egon Albrecht re-equipped with the Me 410 and in June both I./ZG 26 at Königsberg and III./ZG 26 at Fels am Wagram in Austria exchanged their Bf 110s for the *Hornisse*. Previously, on 13 May 1944, II./ZG 26 suffered disastrously. It took off to intercept a formation of B–24 Liberator bombers, near Frankfurt/Oder, but was itself attacked by about 20 P–51 Mustang fighters with disastrous results. Twelve aircraft were lost, the American bomber formation escaped unscathed, and several pilots were killed. Of these, perhaps the most notable was *Obfw* Wolfgang Martin, a Knight's Cross holder. His Me 410 was reported to have rammed an American bomber after ordering his crew to bale out.

Early in May 1943, I./KG 51 *Edelweiss* which had been operating the Ju 88A–4 on the eastern front was withdrawn to Illesheim in Germany for rest and re-equipment with the Me 410. It received its first three Me 410s late in June but it was not until the September 1943 that the unit became operational. It was joined three months later by II./KG 51 which had been formed from V./KG 2.

Both V./KG 2 and I. and II./KG 51 operated by night against the British Isles, providing a reasonably formidable challenge to the defences of RAF Fighter Command. Apart from the conventional bombing sorties carried out by KG 51, a special intruder squadron had been set up within the *Geschwader*. Designated *Fernnachtjäger-Staffel*/KG 51, the unit was commanded by *Maj* Dietrich Puttfarken, a pilot who had previously been awarded the Knight's Cross for bomber operations with KG 51. Puttfarken was probably shot

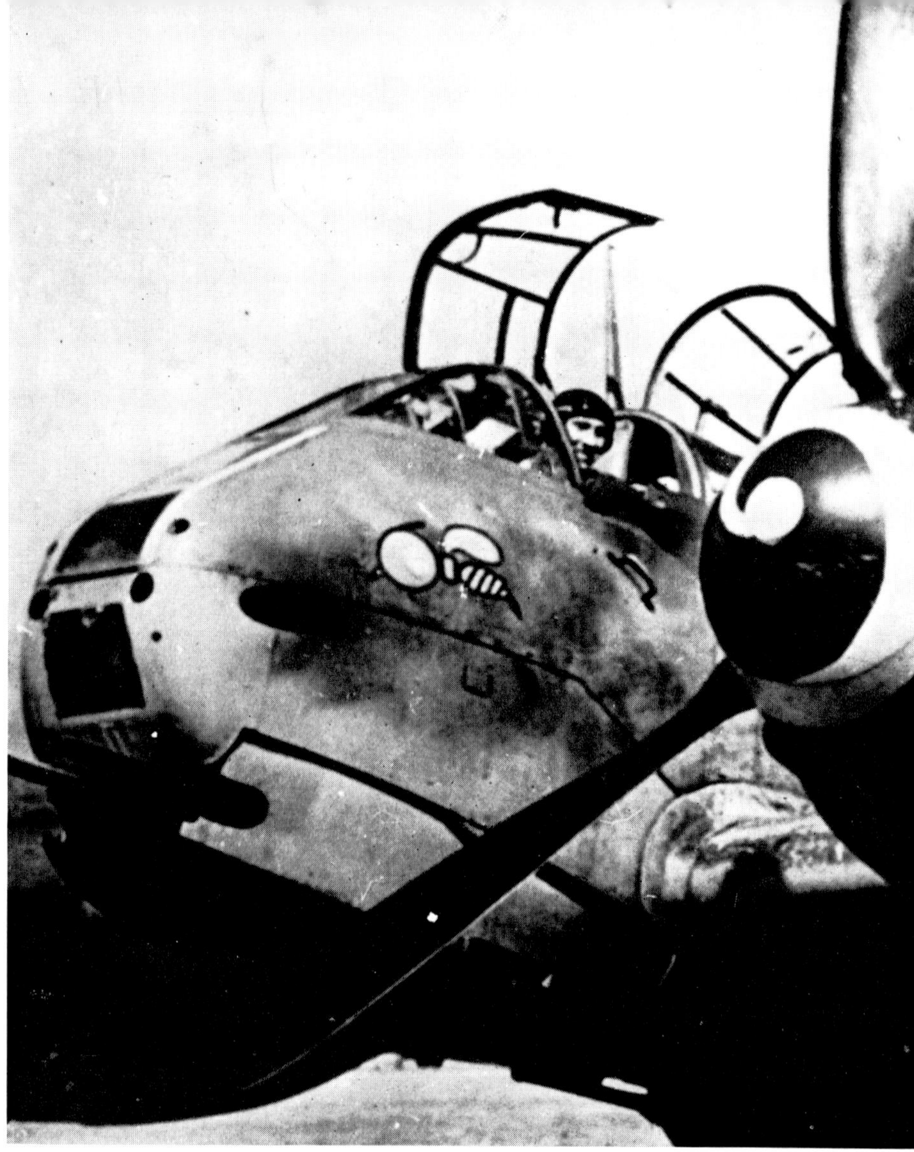

A Me 410B–2/U2 heavy fighter of II./ZG 1 carrying the wasp badge of the unit.

down by an RAF night fighter on 23 April 1944 during a sortie against airfields in the Cambridge area.

Apart from destroyer and bomber operations, the *Hornisse* also began to be delivered in numbers to several *Luftwaffe* long-range reconnaissance units. Apart from 2.(F)/122 in Italy, the aircraft was delivered to 1.(F)/121 in January 1944, 1.(F)/33 in July, 1.(F)/122 in August, 3.(F)/22 in October, 5.(F)/122 in December and 4.(F)/121 in April 1945. The *Hornisse* was also used by several of the *Wetterkundungstaffeln* (meteorological reconnaissance) squadrons including *Wekusta* 51 based at Tours during the Normandy landings.

In April and May 1944, I./NJG 1 and I./NJG 5 both experimented with the Me 410A–2/U2 night fighter, although the type was soon rejected in favour of the Ju 88G and He 219. By July, all seven destroyer *Gruppen* had been re-equipped with the *Hornisse*, these being II./ZG 1 under Albrecht, I., II. and III./ZG 26 under *Oberst* Johann Kogler and I. and III./ZG 76 under *Oberst* Robert Kowalewski.

One of the most notable operations undertaken by I./ZG 76 came on 2 July 1944 when it, in company with Bf 110s from I./ZG 1 and Bf 109s from II./JG 27 claimed the destruction of 45 American aircraft in a tremendous air battle over Budapest. One notable pilot to emerge during this time was *Lt* Rudi Dassow of II./ZG 26 whose Me 410 carried no less than eight 20 mm cannon! Dassow was killed on 25 August 1944 when his aircraft was shot down in flames near Laon.

During the autumn of 1944, the decision was taken to disband all destroyer *Gruppen*, and re-equip them with single-engined fighters. II./ZG 1 became III./JG 76 and later IV./JG 53, I. and II./ZG 26 became I. and II./JG 6, III./ZG 26 formed the basis of III./JG 7 and ZG 76 was incorporated into fighter squadrons. The only other destroyer *Gruppe* to operate the *Hornisse* was IV./ZG 26 *Horst Wessel* which used the machine for a short period in Norway before being re-designated as the new II./JG 5 *Eismeer* in February 1945.

In June 1944, I./KG 51 was withdrawn from operations to be re-equipped with the Me 262A–2a. Three months later, II.*Gruppe* which had operated from Soesterburg, Eindhoven and Gilze Rijen during the Allied invasion of Normandy, was also withdrawn to Germany to exchange its Me 410s for Me 262s. When the war ended, only five reconnaissance squadrons were equipped with the Me 410. These were 2.(F)/22, 4.(F)/121, 1.(F) and 5.(F)/122 with *Luftflotte* 6 and 2.(F)/122 in Italy.

At least one Me 410A–1 was fitted with six rotating 210 mm WGr 21 rocket tubes.

Specification (Me 410A–1)

SPAN	16·40 m	(53 ft 7¾ ins)
LENGTH	12·40 m	(40 ft 8½ ins)
HEIGHT	4·30 m	(14 ft 0½ ins)
EMPTY WEIGHT	6,150 kg	(13,560 lbs)
LOADED WEIGHT	10,650 kg	(23,483 lbs)
MAXIMUM SPEED	625 km/h	(388 mph)
MAXIMUM RANGE	2,330 km	(1,447 miles)

Enzian

DURING THE COURSE of the Second World War, the German aviation industry developed four important ground to air missiles, although none were ever used operationally. Most advanced was the C2 *Wasserfall* produced by the Electromechanische Werke, a supersonic missile based on the Peenemünde series of rockets, of which the V2 was the most notable. Three less sophisticated designs were the supersonic *Rheintochter* and the subsonic Henschel Hs 117 *Schmetterling* and Messerschmitt *Enzian* (Gentian).

The last named missile was designed by Dr Hermann Würster during 1944. Built largely of wood, the *Enzian* was at first glance similar in appearance to the Me 163 *Komet*. The missile was launched from 22-foot long inclined rails mounted on a much-modified 88 mm gun platform, and was to be produced by the Holzbau-Kissing KG at Santhofen, Allgaü.

The *Enzian* E1, E2 and E3 prototype missiles were powered by the Walter RI–210B liquid-fuel rocket engine which developed an initial thrust of 4,400 lbs. This dropped to half by the end of its 70-second burn period so that the missile did not exceed its limiting Mach number of 0·9. The engine used a mixture of SV-*Stoff* (mixed acid) and Br-*Stoff* (petroleum) ignited by a small quantity of T-*Stoff*

Above: *The Messerschmitt Enzian missile mounted on its modified 88 mm anti-aircraft gun platform.*

Right: *An Enzian missile is launched at Peenemünde amidst a cloud of smoke and steam.*

as fuel. Take-off was accomplished with the aid of four Schmidding 109–553 solid-fuel rockets which produced 3,850 lbs thrust each for four seconds.

Two production variants of the *Enzian* were proposed, the E4 with a limiting Mach number of 0·9 and the E5 which it was hoped would attain twice the speed of sound. New rocket motors were designed for the production variants by Dr Konrad and Dr Beck of the *Deutsche Versuchsanstalt für Kraftfahrzeug und Fahrzeugmotoren* (DVK) at Berlin-Dresden. The E4 engine developed similar power to the Walter RI–210B rocket, but the unit for the E5 was to produce a maximum of 5,500 lbs thrust.

The *Enzian* was guided to its target with the aid of a *Knüppel* (joystick) control box, *Kehl* or *Kogge* transmitter and *Strassburg* or *Brigg* superhet receiver. Between April 1944 and January 1945, 38 *Enzians* were launched from Karlshagen, although initial trials were far from satisfactory. At the beginning of 1945, both the *Enzian* and *Rheintochter* were cancelled in favour of the more highly developed Henschel Hs 177 *Schmetterling*, no E5s having been completed.

Specification (Enzian E4)

SPAN	4·00 m	(13 ft $1\frac{1}{2}$ ins)
LENGTH	3·75 m	(12 ft $3\frac{3}{4}$ ins)
LAUNCHING WEIGHT	1,800 kg	(3,960 lbs)
MAXIMUM SPEED	975 km/h	(605 mph)
SPEED AT TARGET	910 km/h	(565 mph)
MAXIMUM RANGE	24·5 km	(15·2 miles)
CEILING	12,500 m	(41,010 ft)

Wartime aircraft projects

P.1101. The P.1101 project was conceived by Messerschmitt early in 1944 as an advanced fighter to make use of the 2,860 lbs thrust Heinkel-Hirth HeS 011 turbojet which was then under development. It was decided to utilise swept wing and tail surfaces in the P.1101, the sweep of the former being adjustable on the ground from 35 to 45 degrees. The turbojet was mounted below and behind the pilot, who was seated well forward above the circular air intake.

Construction of a prototype, the P.1101 V1, was commenced in July at Oberammergau, but despite the boost given to the project by the Emergency Fighter Competition of late 1944, the aircraft was only 80 per cent complete when captured by American forces.

P.1102. During the summer of 1944, Messerschmitt proposed an enlarged version of the P.1101 for unarmed bomber duties. Known as the P.1102, the project

The Messerschmitt P.1101 V1 was only 80 per cent complete when the war ended.

was to be powered by three 1,760 lbs thrust BMW 003 turbojets, one on either side of the nose and one in the rear fuselage, exhausting behind the tail. The aircraft also featured variable wing sweep, the angle of which could be adjusted from 20 to 50 degrees. The undercarriage comprised two tandem wheels below the fuselage balanced by an outrigger at each wing tip.

P.1104. Even before the invasion of Normandy in 1944, the RLM had placed a specification for a rocket propelled target defence fighter with Bachem, Heinkel, Junkers and Messerschmitt. The latter company produced a tiny shoulder wing monoplane design to be built of wood, with the pilot seated just forward of the wing leading edge. It was powered by a 3,750 lbs thrust Walter HWK 509 rocket engine, and utilised a similar dolly and skid ground handling arrangement to that employed by the Me 163. In the event the Bachem Ba 349 gained the production contract and the P.1104 was abandoned.

P.1106. A development of the P.1101, this project was fitted with a butterfly tail.

P.1107. The P.1107 was a proposed heavy bomber design with swept back wings to be powered by four 7,720 lbs thrust BMW 018 turbojets. Two variants were proposed; the P.1107/I in which the engines were slung beneath the wing and the P.1107/II in which the turbojets were completely enclosed inside the wing roots. The second project was to have featured a butterfly tail to avoid any aerodynamic difficulties associated with the efflux from the turbojets.

P.1108. The 1000 × 1000 × 1000 bomber specification called for an aircraft capable of carrying a 1,000 kg (2,200 lb) bomb load over a distance of 1,000 km (621 miles) at a speed of 1,000 km/h (621 mph). Messerschmitt's proposal, the P.1108, was similar in appearance to the P.1107/II design, but was powered by four 2,860 lbs thrust HeS 011 engines.

P.1110. Although employing a very similar wing to the P.1101 project, the Messerschmitt P.1110 introduced entirely new wing and tail surfaces. The initial design featured a jet intake on either side of the fuselage, with a swept back fin and rudder at the base of which was a swept tailplane. The P.1110/II introduced a completely annular intake for its HeS 011 turbojet and a butterfly tail. Armament was proposed as three 30 mm MK 108 cannon in the fuselage with provision for an extra pair of guns in the wing roots.

P.1111. The P.1111 project was a tailless design with the wings swept back at a 45-degree angle to produce a virtual delta planform. The intakes for the 2,860 lbs thrust HeS 011 turbojet were positioned either side of the fuselage in the wing roots, and provision was made for four MK 108 cannon.

P.1112. A development of the P.1111 with additional fuel tankage in the fuselage.

P.1116. A development of the P.1106 project with reduced wing span.

Specifications

PROJECT	SPAN	LENGTH	EMPTY WT	LOAD. WT	MAX SPEED
P.1101	8·24 m 27′ 0½″	9·25 m 30′ 4¼″	2600 kg 5733 lbs	4070 kg 8974 lbs	981 km/h 609 mph
P.1107	17·37 m 57′ 0″	— —	— —	30700 kg 67693 lbs	880 km/h 547 mph
P.1110/I	8·24 m 27′ 0½″	9·25 m 30′ 4¼″	2800 kg 6174 lbs	4295 kg 9470 lbs	1000 km/h 621 mph
P.1111	9·16 m 30′ 0¾″	6·50 m 21′ 4″	2750 kg 6064 lbs	4320 kg 9526 lbs	995 km/h 618 mph

The Avia C.10 series

APART FROM domestic production, a large number of Messerschmitt Bf 109s were built in occupied countries, perhaps the most important of which was Czechoslovakia. Between 1941 and 1944, the country produced 4,147 aircraft, and during the latter year, it was decided that the Bf 109G–14 and Me 262A–1a should be built by the Czech Avia company and several sub-contractors.

In the event, neither type was delivered to the *Luftwaffe* from this source, and after the Czech revolution drove the Germans out of Prague, it was decided to continue production of both types for the national Air Force. All components were collected from the various sub-contractors, inspected for possible sabotage, and delivered to the Avia works at Cakovice. Production of the Bf 109G–14 was then resumed under the designation C.10, the machine being delivered to the Czech Air Force as the S–99.

The C.10 was powered by a 1,475 hp Daimler-Benz DB 605A–1 engine and possessed a similar performance to the standard Bf 109G. An armament of two 13 mm MG 131 machine guns and two 20 mm MG 151/20 cannon were installed. A two-seat trainer version was also produced which was similar to the Bf 109G–12. The aircraft was known as the Avia C.110 or CS–99 with the Czech Air Force. Only a few Me 262A–1a and B–1a variants were completed and it is doubtful whether any were used operationally.

Production of the Avia C.10 was quickly brought to a halt because of the shortage of DB 605 engines. The only other engine available to the industry was the Junkers-Jumo 211F, production of which had been initiated by the Germans before they left. Therefore a C.10 was taken from the production line and re-engined with the 1,300 hp Jumo 211F. The unit drove a large airscrew with three large paddle blades, the excessive torque from which caused the machine to swing violently on take-off. The new variant was known as the Avia C.210 and the S–199 in Air Force service.

An Avia C.210 (otherwise S–199) of the Czech Air Force. The large paddle bladed airscrew used in connection with the Jumo 211 engine is clearly visible.

Only a few S–199s were delivered to the Czech Air Force before a revolution led by Klement Gottwald led to the country joining the Communist camp. The Air Force was quickly re-equipped along Soviet lines, the fighter squadrons receiving the MiG 15 jet fighter. The machine remained in service with the Czech National Security Guard until 1957, when it finally disappeared from strength. The Czech pilots were more than happy to see the back of the '*Mezec*' or Mule as they dubbed their aircraft. A two-seat version of the design was also produced under the alternative designations of Avia C.210C and CS–199.

On 14 May 1948, a favourable vote at the United Nations established the modern state of Israel. On the same day, the new state was attacked by the forces of Egypt, Syria, Lebanon and Iraq, and a desperate attempt was made to create an air defence force. In June 1948 the force became operational, equipped with a small number of ex-RAF Spitfires and Avia C.210s purchased from the Czech government, at 100,000 US dollars each.

The Israeli S–199s often clashed with Egyptian Spitfires, but the determination displayed by Jewish pilots usually resulted in the Arabs coming off worse. In one notable action, an ex-RAF pilot flying a S–199 of the Israeli Defence Air Force, Muddi Alon, shot down two converted Dakotas attempting to bomb Tel Aviv. A few days later he shot down a Spitfire 9, but was mortally wounded. Fighting ceased in January 1949, and the S–199s were exchanged for P–51 Mustangs and de Havilland Mosquitos.

Specification (Avia C.210 or S–199)

SPAN	9·92 m	(32 ft 6½ ins)
LENGTH	8·90 m	(29 ft 2½ ins)
LOADED WEIGHT	3,500 kg	(7,717 lbs)
MAXIMUM SPEED	590 km/h	(367 mph)
NORMAL RANGE	850 km	(528 miles)

Right: *Only a small number of Avia C.210C (CS–199) two-seat trainers were completed.*

Hispano HA–1109

AFTER the successful completion of the Spanish Civil War, Franco's air force decided to institute a new designation system for its motley collection of aircraft. The 45 Bf 109B–2s and 15 Bf 109E–1s were redesignated as the C.4 and C.5 respectively. Ten Bf 109F–3s were purchased by Spain in 1942 under the designation C.10, but then it was decided to undertake licence production of 200 Bf 109G–2s.

Spain's largest aircraft manufacturer, Hispano-Aviacion SA was given the task of producing the aircraft, and to assist in familiarising their engineers, 25 Bf 109G–2s were to be purchased from Germany under the designation C.12. Because of severe bombing, only odd airframe components arrived in Spain, all without engines. In order to get the aircraft into the air, it was decided to substitute the 1,300 hp Hispano HS–12–Z–89 engine for the DB 605.

The aircraft was flown for the first time successfully on 2 March 1945. The installation of the new engine called for some modification to the tail surfaces and a Hamilton airscrew was temporarily fitted. This was to be replaced in the production model, designated Hispano HA–1109–J1L, by an Escher-Wyss propeller of Swiss manufacture. The first machine was delivered to the Spanish Air Force in January 1946, the batch of 25 machines being completed by the beginning of 1947.

Experience proved that the new aircraft was far from successful, and work began on a new variant, the HA–1109–K1L powered by the 1,300 hp Hispano HS–12–Z–17 engine. The machine made its first flight in May 1951, and was followed by the HA–1109–K2L which was equipped with eight 80 mm Oerlikon rockets. The HA–1109–K3L carried rocket armament only, the two 12·7 mm machine guns having been deleted. A twin cannon armed variant was also produced under the designation HA–1112–K1L.

Specification (HA–1109–K1L)

SPAN	9·92 m	(32 ft 6½ ins)
LENGTH	9·10 m	(29 ft 10¼ ins)
HEIGHT	3·40 m	(11 ft 2 ins)
EMPTY WEIGHT	2,220 kg	(4,894 lbs)
LOADED WEIGHT	2,750 kg	(6,063 lbs)
MAXIMUM SPEED	750 km/h	(404 mph)
NORMAL RANGE	600 km	(373 miles)

Above: *Despite being powered by a Hispano HS–12–Z engine, the HA–1109–K1L is clearly related to the Bf 109G.*

Right: *The Hispano HA–1110–K1L was a tandem two-seat trainer version of the HA–1109–K1L.*

Hispano HA–1112 Buchon

DURING the summer of 1953, the eleventh and twelfth production HA–1109–K1L were fitted with the 1,400 hp Rolls-Royce Merlin 500/45 engine under the designation HA–1109–M1L. The HA–1109–M2L was generally similar but carried a 20 mm HS 404 or HS 408 cannon in each wing plus eight 80 mm Oerlikon rockets. The Merlin-engined version was so successful that it was decided to initiate the large-scale production of such a variant under the designation HA–1112–M1L.

The HA–1112–M1L was powered by a 1,400 hp Rolls-Royce Merlin 500/45 engine driving a four-bladed airscrew in an installation somewhat reminiscent of the Spitfire 9. A 20 mm Hispano cannon was mounted in each wing, the barrel protruding well forward of the leading edge, and four 80 mm Oerlikon rockets could be mounted underneath in a staggered arrangement. Eventually 143 HA–1112–M1Ls were converted from HA–1109–K1Ls, the variant entering service with the Spanish Air Force under the designation C.4K. Production of the machine finally ceased in 1958.

In addition to the fighter variants, the Hispano company also built two trainer

Above: *The Hispano HA–1112–M1L was powered by a 1,400 hp Rolls-Royce Merlin engine. It is strange that the last operational Bf 109s were powered by the engine of its arch rival, the Spitfire.*

Right: *Only two prototypes of the HA–1112–M4L two-seat conversion trainer were completed.*

versions of the Bf 109. The HA–1110–K1L was a tandem two-seat trainer conversion of the HA–1109–K1L. The machines' cabin was enclosed by a new clear-blown cockpit hood and no armament was carried. A similar conversion of the Merlin variant was designated HA–1112–M4L but only two aircraft were completed.

Apart from providing the Spanish Air Force with a valuable addition to its equipment from the period that followed the Second World War until well into the 1960s, the HA–1112–M1L also achieved fame as a film star. It was used in the German film *Star of Africa* released in 1959 which told the story of Hans Marseille, but it is perhaps more recently that the HA–1112–M1L became familiar to British film audiences. Twenty-six machines were acquired from the Spanish government and used in the United Artists' film *Battle of Britain* which was released on 15 September 1969. Painted in *Luftwaffe* colours of the 1940 period, the Bf 109 again clashed with the Spitfire, albeit in a modified form, for perhaps the last time.

Specification (HA–1112–M1L)

SPAN	9·92 m	(32 ft 6½ ins)
LENGTH	9·10 m	(29 ft 10½ ins)
HEIGHT	3·40 m	(11 ft 2 ins)
EMPTY WEIGHT	2,660 kg	(5,865 lbs)
LOADED WEIGHT	3,180 kg	(7,011 lbs)
MAXIMUM SPEED	675 km/h	(419 mph)
NORMAL RANGE	765 km	(475 miles)

Right: *A standard production Hispano HA–100–E1 trainer of the Spanish Air Force.*

Hispano HA–100–E1

AFTER THE END of the Second World War, Prof Willi Messerschmitt joined the Spanish Hispano Aviacion company, supervising the design and construction of a small tandem two-seat trainer. Designated HA–100–E1, the aircraft was a neat but angular low-wing monoplane with a bubble cockpit and a retractable nosewheel undercarriage. The prototype flew for the first time on 10 December 1954 powered by a 750 hp ENMA Beta B4 radial engine. It carried an armament of two 12·7 mm machine guns and a 450 lb load could be mounted underwing.

The prototype was followed by a second similar machine, but the third and fourth machines were fitted with a 800 hp Wright 957 C–7B–A1 radial under the designation HA–100–F1. There was little to choose between the two variants, but the elastic mounting of the Wright engine proved unpopular with Spanish pilots and the HA–100–E1 was chosen for production under the Air Force designation E.12.

Production aircraft differed in having the engine moved further forward and the provision of a large dorsal fin. Forty aircraft were ordered for the Spanish Air Force, and consideration was given at one stage for the production of a German variant for the *Luftwaffe* under the designation Messerschmitt Me 100. The machine supplemented and eventually replaced the North American T–6 Texan as the Spanish Air Force's advanced trainer.

Specification (HA–100–E1)

SPAN	10.40 m	(34 ft 1½ ins)
LENGTH	8·95 m	(29 ft 4½ ins)
HEIGHT	3·00 m	(9 ft 10 ins)
EMPTY WEIGHT	1,970 kg	(4,343 lbs)
LOADED WEIGHT	2,930 kg	(6,459 lbs)
MAXIMUM SPEED	440 km/h	(274 mph)
NORMAL RANGE	1,320 km	(820 miles)

The Hispano HA–200–R1 twin-engined jet trainer was also known as the Messerschmitt Me 200.

Hispano HA–200–R1 Saeta

ONE OF the most interesting aircraft produced by the Hispano company was the HA–200–R1 *Saeta* (Arrow) designed by Prof Messerschmitt. The design, which bore the Messerschmitt name Me 200 *Pfeil*, was the first jet aircraft to be built in Spain. It was a neat low-wing monoplane somewhat similar in appearance to the HA–100–E1, but powered by two 880 lbs thrust Turboméca Marboré II turbojets mounted side by side within the fuselage.

The prototype HA–200–R1 first flew on 16 August 1955 and a small pre-production batch was produced for use by the Spanish Air Force. This variant differed in having a pressurised cabin and could carry four 22 lb rockets and four 110 lb bombs. In the event the HA–200–R1, or E.14 as it was to be designated in Spanish service, was never built in numbers. A variant of the basic design was also proposed, the HA–231–R1 *Guión* which was to be built in two forms. The first was a side-by-side two-seat trainer and the second was a six-seat liaison aircraft with three pairs of seats.

Specification (HA–200–R1)

SPAN	10·42 m	(34 ft 2½ ins)
LENGTH	8·90 m	(29 ft 2½ ins)
HEIGHT	3·25 m	(10 ft 8 ins)
EMPTY WEIGHT	1,680 kg	(3,704 lbs)
LOADED WEIGHT	2,640 kg	(5,821 lbs)
MAXIMUM SPEED	700 km/h	(435 mph)
NORMAL RANGE	1,610 km	(1,056 miles)

Some notes on aircraft and unit designations

EARLY AIRCRAFT produced by the Harth-Messerschmitt partnership bore the designation 'S' (for Segelflugzeuge or sailplane) and a progressive number. After the resignation of Harth, his name was dropped from the company's title and the abbreviation 'M' (for Messerschmitt) was introduced to replace the previous letter.

This system was retained until the design of the M 37 tourer in 1933. Following the advent of the *Nazis* to power in 1933, the German Air Ministry rationalised all aircraft designation systems, each design being allocated a number from a single progressive list, prefixed by '8'. Therefore, the Messerschmitt M 37 was now described as the 8–108 in Air Ministry records.

The progressive numbers were allocated in batches, for example 133 to 134 was used by Bücker, 135 to 144 by Blohm und Voss and 145 to 150 by Gotha. Each number was prefixed by a two or three letter code indicating the manufacturer of the aircraft. For example, Arado used 'Ar', Heinkel used 'He' and Henschel used 'Hs'. All Messerschmitt designs at this time, which were of course built by the Bayerische-Flugzeugwerke, carried the abbreviation 'Bf'. Following the renaming of the company as the Messerschmitt AG on 11 July 1938, the abbreviation 'Me' was introduced, but previous designs retained the 'Bf' designation in all *official* German publications.

The aircraft designation was followed by a letter which identified a major sub-type; e.g.: the Bf 109E, F, G, etc. For more minor modifications a number was used; e.g.: Me 410A–1, A–2, A–3, etc. Finally, very small modifications were identified by the letter 'U' and a number; e.g.: Bf 110C–2/U1. Conversion packs were recognised by the addition of the letter 'R' and a number; e.g.: Bf 109G–6/R1, and the abbreviation 'Trop' indicated that the variant was fitted with tropical equipment.

All prototypes carried a *Veruschs* (experimental) number, the first prototype of the Messerschmitt *Gigant* being the Me 321 V1, the second the Me 321 V2 and so on. Pre-production aircraft carried the normal suffix letter and the number '0'. Therefore, the Me 163B–0 was the pre-production aircraft for the 'B' series.

The basic *Luftwaffe* unit was the *Staffel* or squadron which was normally made up of between nine and twelve aircraft. Three or four *Staffeln* were combined to form a *Gruppe* and three or four *Gruppen* were combined to form a *Geschwader* or wing. *Geschwader* were identified by an Arabic number, *Gruppen* within a *Geschwader* by a Roman numeral. Independent *Gruppen* were identified by an Arabic number as were all *Staffeln*. Each type of unit was prefaced by an abbreviation which identified its rôle. For example, the letters 'JG' identified a *Jagdgeschwader* or fighter wing. The designation II./JG 52 identified the second group of fighter wing 52, while the designation 2./JG 52 identified the second squadron of the same wing.

Abbreviations

Aufkl.Gr	*Aufklärungsgruppe* Reconnaissance group
DFS	*Deutsches Forschungsinstitut für Segelflug* German research institute for sailplanes
DLV	*Deutscher Luftsport-Verband* German sport flying association
DVL	*Deutsche Versuchsanstalt für Luftfahrt* German experimental institute for aviation
DVS	*Deutsche Verkehrsfliegerschule* German commercial pilot school
(F)	*Fern-* Long-range
(H)	*Heeres-* Army co-operation
JG	*Jagdgeschwader* Fighter wing
KG	*Kampfgeschwader* Bomber wing
LG	*Lehrgeschwader* Instructional wing
(N)	*Nah-* Short-range
NJG	*Nachtjagdgeschwader* Night fighter wing
Ob.d.L	*Oberbefelshaber der Luftwaffe* High command of the Luftwaffe
RLM	*Reichsluftfahrtministerium* German aviation ministry
SG	*Schlachtgeschwader* Ground attack wing
St.G	*Stukageschwader* Dive bomber wing
TG	*Transportgeschwader* Transport wing
ZG	*Zerstörergeschwader* Destroyer wing

Luftwaffe commissioned rank abbreviations

Lt	*Leutnant*	(Pilot Officer)
Oblt	*Oberleutnant*	(Flying Officer)
Hptm	*Hauptmann*	(Flight Lieutenant)
Maj	*Major*	(Squadron Leader)
Oberstlt	*Oberstleutnant*	(Wing Commander)
Oberst	*Oberst*	(Group Captain)
Gen Maj	*General-Major*	(Air Commodore)
Gen Lt	*General-Leutnant*	(Air Vice Marshal)
Gen d Fl	*General der Flieger*	(Air Marshal)
Gen Oberst	*General-Oberst*	(Air Chief Marshal)
Gen Feldm	*General-Feldmarschall*	(Marshal of the Royal Air Force)

Type and Production List

The list that follows includes all aircraft allocated a Messerschmitt or RLM type number, but excludes Messerschmitt projects. In addition, foreign-built Messerschmitt aircraft are also listed. The section on power plants includes the number of engines fitted, the horsepower or pounds thrust and the type.

TYPE NO.	AIRCRAFT TYPE	ENGINE	BUILT FOR (OWNER)	DATE	NO. BUILT
S 9	Glider	—	—	1920	1
S 10	Glider	—	—	1921	1
S 11	Glider	—	—	1921	1
S 12	Glider	—	—	1921	1
S 13	Glider	—	—	1921	1
S 14	Glider	—	—	1922	2
S 15	Powered glider	1 × 14 Douglas	—	1924	1
S 16	Powered glider	1 × 21 Douglas Sprite	—	1924	2
M 17	Lightplane	1 × 32 Bristol Cherub	Theo Cronweiss	1925	
M 18	Light airliner	1 × 125 Siemens Sh 12	Nordbayerische Verkehrsflug	1926	26
M 19	Lightplane	1 × 24 ABC Scorpion	DVL	1927	2
BFW U 12b	Biplane trainer	1 × 108 Siemens Sh 12	DVS and private pilots	1926	
BFW 1	Biplane trainer	1 × 108 Siemens Sh 12	—	1926	
BFW 3	Biplane trainer	1 × 112 Siemens Sh 13	—	1927	
M 20	Light airliner	1 × 500 BMW VIa	*Lufthansa*	1928	15
M 21	Biplane trainer	1 × 125 Siemens Sh 12	Verkehrsministerium	1928	1

TYPE NO.	AIRCRAFT TYPE	ENGINE	BUILT FOR (OWNER)	DATE	NO. BUILT
M 22	Biplane bomber	2 × 500 Siemens Jupiter	Verkehrsministerium	1928	1
M 23	Lightplane	1 × 80 Siemens Sh 11	private use and racing	1928/30	
M 24	Light airliner	1 × 320 BMW Va	Nordbayerische Verkehrsflug	1929/30	4 or 5
M 25	Special project	—	Ernst Udet	1930	None
M 26	Light airliner	1 × 100 Siemens Sh 11	—	1930	1
M 27	Lightplane	1 × 110 Siemens Sh 12	private use and racing	1930	
M 28	Mailplane	1 × 525 BMW Hornet	*Lufthansa*	1932	2
M 29	Racing lightplane	1 × 150 Argus As 8	racing use	1932	4
M 30	Light airliner project	1 × 175 Wright Whirlwind	—	1931	None
M 31	Lightplane	1 × 60 BMW Xa	private use	1932	1
M 32	Biplane trainer project	—	Verkehreministerium	1932	None
M 33	People's aircraft	1 × 15 DKW	private use	1932	Mock-up only
M 34	Project	—	—	1933	None
M 35	Lightplane	1 × 150 Siemens Sh 14a	private use	1933	
M 36	Light airliner	1 × 380 A.S.Serval	LARES	1934	1
Bf 108	Tourer	1 × 240 Argus As 10C	private use and *Luftwaffe*	1934/45	885
Bf 109	Fighter	1 × 1300 Daimler-Benz DB 601F	*Luftwaffe*	1935/45	about 34,000
Bf 110	Destroyer	2 × 1300 Daimler-Benz DB 601F	*Luftwaffe*	1936/45	3,522
Me 155	High altitude fighter	1 × 1500 Daimler-Benz DB 628	*Luftwaffe*	1943	None (3 Bv 155s)

TYPE NO.	AIRCRAFT TYPE	ENGINE	BUILT FOR (OWNER)	DATE	NO. BUILT
Bf 161	Reconnaissance aircraft	2 × 960 Daimler-Benz DB 600	*Luftwaffe*	1938	2
Bf 162	Light bomber	2 × 960 Daimler-Benz DB 600	*Luftwaffe*	1937	3
Bf 163	STOL Recce aircraft	1 × 240 Argus As 10C	*Luftwaffe*	1938	1
Me 163	Rocket fighter	1 × 3750 Walter HWK 509A	*Luftwaffe*	1941	364
MeC 164	Airliner project	2 × 575 Argus As 411	*Lufthansa*	1941	None
Me 208	Tourer	1 × 240 Argus As 10C	private use and *Luftwaffe*	1941	2
Me 209 V	Experimental fighter	1 × 1200 Daimler-Benz DB 601	*Luftwaffe*	1938	4
Me 209A	Fighter	1 × 1750 Junkers Jumo 213	*Luftwaffe*	1943	4
Me 210	Fighter-bomber	2 × 1395 Daimler-Benz DB 601	*Luftwaffe*	1939	352
Me 261	Reconnaissance aircraft	2 × 2950 Daimler-Benz DB 610	Olympic Games (Tokyo)	1940	3
Me 262	Jet fighter	2 × 1980 Junkers Jumo 004B	*Luftwaffe*	1942	1,433
Me 263	Rocket fighter	1 × 4430 Walter HWK 509C	*Luftwaffe*	1944	1
Me 264	Heavy bomber	4 × 1340 Junkers Jumo 211	*Luftwaffe*	1942	2
Me 265	Fighter project	2 × 1750 Daimler-Benz DB 603	*Luftwaffe*	1942	None
Me 309	Fighter	1 × 1750 Daimler-Benz DB 603	*Luftwaffe*	1942	4
Me 310	High altitude project	2 × 1750 Daimler-Benz DB 603	*Luftwaffe*	1942	None
Me 321	Glider transport	—	*Luftwaffe*	1941	175
Me 323	Heavy transport	6 × 990 Gnôme-Rhône 14	*Luftwaffe*	1941	201
Me 328	Fighter-bomber	2 × 660 Argus As 014	*Luftwaffe*	1943	

TYPE NO.	AIRCRAFT TYPE	ENGINE	BUILT FOR (OWNER)	DATE	NO. BUILT
Me 329	Fighter-bomber project	2 × 1750 Daimler-Benz DB 603	*Luftwaffe*	1943	None
Me 409	Fighter project	1 × 1750 Daimler-Benz DB 603	*Luftwaffe*	1942	None
Me 410	Fighter-bomber	2 × 1750 Daimler-Benz DB 603	*Luftwaffe*	1944	1,126
Me 609	Fighter-bomber project	2 × 2000 Daimler-Benz DB 603	*Luftwaffe*	1944	None
Nord 1000	Communications aircraft	1 × 240 Argus As 10C	French navy and air force	1945	285
Nord 1001	Communications aircraft	1 × 240 Renault 6Q–10	French navy and air force	1946	
Nord 1002	Communications aircraft	1 × 240 Renault 6Q–11	French navy and air force	1946	
Nord 1003	Communications project	1 × 240 Potez 6D	French navy and air force	1946	None
Nord 1004	Communications project	1 × 305 Potez	French navy and air force	1946	None
Avia C.10	Fighter	1 × 1475 Daimler-Benz DB 605	Czech Air force	1945	
Avia C.110	Fighter-trainer	1 × 1475 Daimler-Benz DB 605	Czech Air Force	1945	
Avia C.210	Fighter	1 × 1300 Junkers Jumo 211	Czech and Israeli Air Forces	1947	
HA–1109	Fighter	1 × 1300 Hispano HS–12–Z–17	Spanish Air Force	1946	225
HA–1110	Fighter-trainer	1 × 1300 Hispano HS–12–Z–17	Spanish Air Force	1947	
HA–1112	Fighter	1 × 1400 Rolls-Royce Merlin	Spanish Air Force	1953	143 (conversions)
HA–100–E1	Trainer	1 × 750 ENMA B4 Beta	Spanish Air Force	1954	42
HA–200–R1	Jet trainer	2 × 880 Turbomeca Marbore II	Spanish Air Force	1956	